Cataclysmic Cosmic Events
and How to Observe Them

For other titles published in this series, go to
www.springer.com/series/5338

Martin Mobberley

Cataclysmic Cosmic Events and How to Observe Them

 Springer

Martin Mobberley
Suffolk, UK
e-mail: martin.mobberley@btinternet.com

ISBN: 978-0-387-79945-2   e-ISBN: 978-0-387-79946-9
DOI: 10.1007/978-0-387-79946-9

Library of Congress Control Number: 2008920267

Printed on acid-free paper

springer.com

Acknowledgements

As was the case with my five previous Springer books, I am indebted to my fellow amateurs who have generously donated pictures of themselves, and images of their results, to this project. I am especially indebted to Seiichiro Kiyota for his considerable help in contacting Japanese nova patrollers who donated images of themselves for Chapter 2. In addition, his countrymen Osamu Ohshima and Tomohisa Ohno kindly assisted in obtaining photographs of the legendary discoverer Minoru Honda. In alphabetical order I would like to sincerely thank all of the following astronomers without whom this book would not have been possible:

Ron Arbour, Mark Armstrong, Tom Boles, Denis Buczynski, John Fletcher, Katsumi Haseda, Guy Hurst, Ken Kennedy, Seiichiro Kiyota, Robin Leadbeater, Gianluca Masi, Yuji Nakamura, Hideo Nishimura, Tomohisa Ohno, Osamu Ohshima, Arto Oksanen, Gary and Jean Poyner, Yukio Sakurai, John Saxton, Jeremy Shears, Mike Simonsen, Kesao Takamizawa, Akira Takao, Dave Tyler, Mauri Valtonen, and Minoru Wakuda.

Thanks also to the following organizations: the British Astronomical Association (BAA), the European Space Agency (ESA), the National Aeronautics and Space Administration (NASA), the Space Telescope Science Institute (STScI), and the Solar and Heliospheric Observatory (SOHO).

I would also like to thank all at Springer who have made this book possible, in particular, Jenny Wolkowicki, who has conscientiously guided this book and my previous three Springer books through the complex production process with an essential injection of wry humor, despite being remorselessly "chased by alligators"! Without Jenny none of these books would exist.

Finally, I am indebted to my father Denys Mobberley for his support in all my astronomical activities during my past forty years in this hobby.

Contents

About the Author

Martin Mobberley is a well-known British amateur astronomer who images a wide variety of objects, including comets, planets, novae, supernovae, and asteroids (including 7239 Mobberley). He has written five previous amateur and practical astronomy books for Springer as well as three children's books about astronomy and space travel. Martin has served as the British Astronomical Association's president and received the BAA's Goodacre award in 2000. He is a regular contributor to the British magazine *Astronomy Now* and the *BBC Sky at Night* publication and has appeared as a guest on Sir Patrick Moore's "Sky at Night" TV program on numerous occasions since the late 1990s.

Preface

In the Victorian era – or for non-British readers, the mid-to-late nineteenth century – amateur astronomy tended to center on Solar System objects. The Moon and planets, as well as bright comets, were the key objects of interest. The brighter variable stars were monitored, but photography was in its infancy and digital imaging lay a century in the future.

Today, at the start of the twenty-first century, amateurs are better equipped than any professionals of the mid-twentieth century, let alone the nineteenth. An amateur equipped with a 30-cm telescope and a CCD camera can easily image objects below magnitude 20 and, from very dark sites, 22 or 23. Such limits would have been within the realm of the 100- and 200-inch reflectors on Mount Wilson and Mount Palomar in the 1950s, but no other observatories. However, even those telescopes took hours to reach such limits, and then the photographic plates had to be developed, fixed, and examined by eye. In the modern era digital images can be obtained in minutes and analyzed 'on the fly' while more images are being downloaded. Developments can be e-mailed to other interested amateurs in real time, during an observing session, so that when a cataclysmic event takes place amateurs worldwide know about it. As recently as the 1980s, even professional astronomers could only dream of such instantaneous communication and processing ability.

This new-found power in the hands of backyard observers has enabled them to reach outside the Solar System, and outside our own galaxy, and permitted them to monitor colossal outbursts and explosions taking place as far out as the edge of the observable universe. Many of these events can only be described as cataclysmic and involve violent outbursts of energy ranging from 10^{20} to 10^{47} joules. A joule is a measure of energy and a watt is a measure of power, that is, a watt = a joule per second. When describing extraordinarily energetic events it is common to use the terms joules, or ergs, to describe the total amount of energy released in a single event, whether that event is a nova or a supernova exploding, or a massive solar flare. One joule is equivalent to 10 million ergs. An erg has been compared to the amount of energy needed by a mosquito to achieve take off! A 1-megaton hydrogen-bomb going off produces something like 4×10^{15} joules. The famous volcanic explosion of Krakatoa is thought to have produced 5×10^{17} joules. But beyond 10^{20} joules we start moving into distinctly astronomical territory.

The mildest events discussed in this book are the solar flares emitted by our own Sun that range in energy output from 10^{20} joules, for the barely detectable events, to the most violent flares, registering between 10^{25} and 10^{26} joules. Flares observed on the nearest flare stars to our Solar System can routinely be of this order of magnitude or greater.

Monitoring cataclysmic variable stars (CVs) has become an all-consuming hobby for many dedicated amateurs in the last few decades. Such binary stars can flare dramatically in brightness by several 100-fold when in outburst, but unlike novae or supernovae they can do it all over again months later. In addition they can be seen varying in brightness over the course of mere minutes at the eyepiece. Such stars rapidly become like old but highly eccentric friends to the CV specialist.

The more violent cousins of CVs, novae, and recurrent novae are highly prized targets for the dedicated patrollers whose jobs have become far less messy than when all patrol films needed developing in noxious chemicals. In a full-blown nova outburst somewhere between 10^{37} and 10^{38} joules of energy are released. This is comparable with the amount of energy our own Sun releases in 1000 to 10,000 years!

Moving up in energy levels we next come to the supernovae and hypernovae that well-equipped amateur astronomers have become so proficient at detecting in the last 10 years. Here we add even more zeroes as their outbursts lie within the 10^{44}–10^{46}-joule region – like, say, 100 million novae going off! Supernovae have become a hot topic in recent times, and the study of the behavior of the ultrabright Type Ia supernova light-curves have huge implications for the expansion rate of the universe.

The final two topics covered in this book are truly mind-blowing in terms of their violence and energy output. Active galaxies, containing supermassive black holes at their centers, can swallow a solar mass every month, releasing huge amounts of energy that amateur astronomers can observe as a fluctuation in light output. The luminosity of some of these objects is truly mind-boggling, enabling them to be seen in modest aperture telescopes every clear, dark night, despite being billions of light-years away. Unlike novae and supernovae, active galaxies, or rather their nuclei, are not one-off explosions. They fluctuate in brightness, yes, but it is their constant power output, year after year, decade after decade, that is impressive. Their luminosity is staggering, in excess of 10^{40} watts in the cases of the most extreme quasars/quasi-stellar objects (QSOs), or tens of trillions the power output of our own Sun.

Active galaxies and supernovae are, arguably, only challenged in violence by one category of object, gamma-ray bursts, or 'GRBs.' Once again, black holes are thought to be involved, and such is the brilliance of these transient events that objects near the edge of the detectable Universe would, if you were looking in the right place, at the right time, be visible in binoculars! The active galaxies and GRBs currently battle for the record of the most distant, highest redshift objects known; the former shine at us across billions of light-years due to the radiation from the nucleus of a galaxy surrounding a billion-solar-mass black hole. The latter, in the most violent cases, shine briefly but violently, the result of a single supermassive star disappearing from our universe into a black hole of its own making.

Remarkable though these events are, simply in terms of joules and watts, the truly amazing quality of them is that they can all be monitored from a backyard observatory, by any amateur with a decent aperture telescope and modern CCD equipment. In many cases they can be monitored visually, too.

Amateur astronomers have always sought to push back the boundaries and contribute to real science. As space probes have traveled to every planet in our Solar System professional astronomers have turned their attentions further afield: to cataclysmic stars and novae in our own galaxy, to supernovae in external galaxies, and to active galaxies and GRBs. The keenest amateurs have not been daunted by these new challenges and have followed the professionals' example. If you are a backyard observer and fancy some serious cosmological challenges within our galaxy, outside it, and out to the edge of the observable universe, this book should (we hope!) help you on your way, along with the Internet, which is a constant source of star charts and images for the hundreds of high-energy objects worthy of detailed scrutiny.

Martin Mobberley
January 2008

Cataclysmic Variables

In the last 20 years, with increasing numbers of amateur astronomers owning large-aperture telescopes and sensitive Charge-coupled device (CCD) detectors, there has been a major increase in the popularity of observing challenging, but dramatic, variable stars. Some so-called 'eruptive' stars easily qualify as dramatic even without a telescope. Brilliant naked-eye supernovae (within our own galaxy) have occurred in historic times, and some novae have also become easy naked-eye objects. However, the term cataclysmic variable, or CV, is generally used to describe rather fainter telescopic objects that, nevertheless, have a dramatic outburst magnitude compared to their normal state. Quite often these stars are invisible visually, at minimum, even in large amateur telescopes. Indeed, they are often invisible in amateur CCD images, too, except when in outburst. However, the brightest examples are easy binocular objects.

The CVs we are considering here are eruptive systems comprised of binary stars, where material flowing from the secondary star onto an accretion disk causes the system to outburst in a spectacular fashion; the object may brighten by 100-fold (5 magnitudes) in the course of 1 day. In addition, unlike novae (see Chapter 2), CVs can have outbursts on a regular basis, and a keen observer may witness many outbursts over a period of years. Understanding the accretion disk of a CV is critical to understanding how these objects can vary so dramatically. Strictly speaking, the term CV embraces many objects that have a cataclysmic outburst, including supernovae and novae. However, the type of CV we are concentrating on in this chapter is the dwarf nova, that is, a binary star system with a smaller outburst range than a classical nova, but which outbursts more than once. In practice the keenest observers of dwarf novae tend to loosely interchange the term with CV. In the online era typing two letters is a lot quicker than typing 9 or 10.

A Binary System

A typical CV consists of a white dwarf star (the primary) and a red dwarf star (the secondary) orbiting about their common center of gravity very rapidly (Figure 1.1). A white dwarf star cannot have a mass greater than 1.4 times the mass of our Sun. This mass restriction is called the Chandrasekhar limit. Essentially, a white dwarf is the ash from the controlled nuclear fusion bomb that was once a healthy star. In happier times the nuclear forces and the star's gravity existed at a stable equilibrium, but as the hydrogen ran out the star expanded, its outer layers drifted away, and the ash from the furnace was all that remained: helium, carbon, oxygen, and all the heavier elements. This core is only as big as Earth, but with a mass of up to 1.4 solar masses the density is colossal, typically

M. Mobberley, *Cataclysmic Cosmic Events and How to Observe Them*,
DOI: 10.1007/978-0-387-79946-9_1, © Springer Science+Business Media, LLC 2009

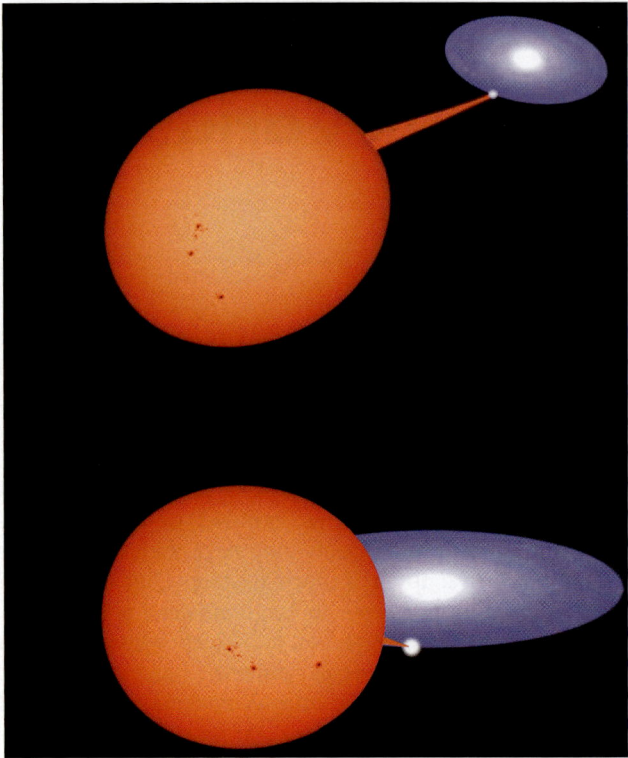

Figure 1.1. A cataclysmic variable usually consists of a red star nearing the end of its life and a compact white dwarf surrounded by an accretion disk. The lower image shows that if our viewing angle is close to the orbital plane of the system then eclipses of the various components can be witnessed. Graphic by the author.

100,000 times the density of Earth. At this point the concept of a black hole may spring to mind, but we are not yet talking of densities quite as high as that. A property of quantum mechanics prevents the white dwarf collapsing any further than 'Earth-size' at this point. This property, the Pauli Exclusion Principle, prevents any two electrons from occupying the same space at the same time, and provided the white dwarf's mass is below about 1.4 solar masses, Pauli wins. However, once the Chandrasekhar limit is exceeded the gravitational forces will be too great, and the star will collapse to become one big atomic particle, that is, a neutron star just 10 or 20 km in diameter.

The current thinking is that for stars of more than 2 or 3 solar masses (the Tolman–Oppenheimer–Volkoff limit) the neutron star may be scrunched further into a quark star. Beyond 5 solar masses a black hole results, with an escape velocity greater than that of the speed of light. However, we have digressed here. Essentially, a white dwarf is a burnt-out star with a maximum mass of 1.4 solar masses and a typical mass of 0.6 solar masses. It is very hot when formed, but since it has no internal energy, it will gradually radiate away its heat and cool down. This is the source of a white dwarf's faint luminosity that initially will exhibit a high color temperature (blue-white) but gradually fade and redden. In practice most observed white dwarfs have a temperature of between 8000 K and 40,000 K, although 4000–150,000 K has also been observed. Astronomers currently doubt whether the age of the universe (13.6 billion years) is long enough for a cold black dwarf to form. In absolute magnitude terms white dwarfs are dim. Perhaps the best-known white dwarf is Sirius B, the companion of the brightest star in the night sky, Sirius A. Sirius B has a mass similar to our Sun, a diameter slightly smaller than our Earth, and is only 8.6 light-years away. Yet, despite its proximity, it only has a magnitude of 8.44, and an absolute magnitude (brightness if it were placed at 32.6 light-years distant) of 11.4, some 400 times dimmer than our own Sun.

The red dwarf component of a CV is rather less bizarre. Typically much less than half the mass of our own Sun, a red dwarf is lighter but physically larger than a white dwarf, with a relatively cool surface temperature of around 3000 K. The white dwarf may look tiny in comparison, but it is heavier and much denser. Like white dwarfs, red dwarfs are rather feeble stars, and their low masses and low core temperatures ensure they will have a long, if boring, life. However, if they are orbiting a superdense white dwarf, maybe 'boring' is not the right term! Some red dwarfs have luminosities as low as a 10,000th of that of our own Sun.

Stars Almost Touching

At this point it is worth stressing just how close the red and white dwarfs are to each other and, in particular, how ludicrously rapidly they are orbiting one another. Our Sun has a diameter of 1.39 million km, and we complete an orbit around it, every year, at a mean distance of 149.6 million km. Even the closest planet to the Sun, Mercury, takes 88 days to orbit our star, at a mean distance of 58 million km. In contrast, CV orbital periods range from 78 minutes (this is not a typographical error!) to about 12 hours. Extraordinary! There is an orbital period gap of between 2 hours and 3 hours (more on this later), but essentially, CV components orbit each other in periods of hours, not days, months, or years. Of course, this naturally follows from two massive bodies being in such close contact. The square of the orbital period of a planet around a star is proportional to the cube of its distance from the Sun (Kepler's third law). Thus, Earth orbits our Sun in 1 year, but Saturn, 9.4 times further from the Sun, orbits in slightly more than 29 years. Put another way, the orbital period is proportional to $\sqrt{(\text{solardistance})^3}$. With two bodies of similar mass the two stars comprising a CV rotate about a center of gravity that is not at the center of the biggest star, but, you get the point: closer means faster.

The typical separation distance between the white and red dwarf components of a CV is similar to the radius of our own Sun! A CV system, purely in terms of size, might look like a multi-Jupiter-sized object (the red dwarf) and an Earth-sized object (the white dwarf) spinning about their common center of gravity, and separated by less than double the Earth–Moon distance. Yet when a dwarf nova outburst occurs, this relatively diminutive arrangement can go from being an object with, say, a luminosity of only a few percent of the brightness of our own Sun to one that would equal it, or even outshine it by 10 or 20 times. (As we shall see later, nova outbursts are considerably brighter and more violent events than their dwarf nova counterparts.) The hot (typically >12,000 K) white dwarf in a CV system is relatively bright and emits most of its energy in the ultraviolet region of the spectrum. Conversely, the dim red dwarf (at approx. 3000 K) peaks in output in the infrared. However, the brightest part of a CV binary system in outburst is the accretion disk hot spot, which we will examine shortly.

The underlying reason that CV outbursts are so spectacular is that hydrogen is flowing from the red dwarf to the superdense white dwarf. The outbursts occur not at the surface of the white dwarf but on its accretion disk. Why should material flow from one star to another anyway, you may ask? Surely one would expect each star to hold onto its own material, unless the two stars were almost touching? The answer to this is that the two stars are incredibly close to each other, and the tenuous gas at the outer edge of the red dwarf's gravitational influence can cross the boundary into the superdense white dwarf's domain. The red dwarf still functions as a low-output star, and the pressure of its atmosphere can push material beyond the so-called Roche lobe (after the nineteenth-century mathematician Edouard Roche), the pear-shaped region within which a star can just hold onto its own material. The white dwarf has a Roche lobe, too, and where these pear-shaped regions meet, at the L1 Lagrangian point, we find the easiest place for hydrogen, pushed by the red dwarf's atmospheric pressure, to fall toward the denser and heavier primary star. It is analogous to filling a ditch with too much water. Eventually, water will slop over the top and flow down into the neighboring ditch at the lowest point between the two.

Red dwarfs in CV systems that have overflowed their Roche lobe are invariably tidally locked, like our Moon. They orbit the white dwarf in the same period as they rotate, the transferred hydrogen always taking the same path. Because the secondary is orbiting the primary at great speed the hydrogen does not fall straight onto the surface of the hot superdense primary. Instead, it swings around the white dwarf in a complex series of ellipses, colliding with its own stream every orbit, but eventually settling down into a ring around the white dwarf as angular momentum, collisions with existing material, and gravitational forces come to an agreement. With a constant stream of material flowing from secondary to primary an accretion disk builds up around the white dwarf, and, over time, the material slowly spirals down onto the surface of the superdense primary itself.

At the point where the material from the secondary hits the accretion disk we get a hot spot (sometimes called a bright spot). The visual behavior of CVs, as observed by amateur astronomers, is inextricably linked with accretion disk phenomena. The more information gathered by amateur observers the more the professionals can tweak their computer simulations and work out what is happening. It might, at first, be thought that there could not be much to learn about CVs after half a century of analysis. However, more is being learned every year, especially as no two CVs are the same, and it is only in recent years that larger amateur telescopes, Go To drives, and CCD cameras have enabled such extensive amateur coverage. The outburst periods of CVs vary considerably. Some outburst regularly and predictably, others are highly unpredictable. Spotting the first signs of an outburst from a rarely visible CV carries a certain amount of kudos in the variable star community.

Accretion Disks, Hot Spots, and Eclipses

Cataclysmic variables are so named for good reason: their outbursts warrant the term cataclysmic, even though the star is not destroyed in the outburst. To distinguish these outbursts from the even more spectacular, but rarer, nova outbursts, they are sometimes called, as we saw earlier, dwarf nova outbursts. A dwarf nova/CV outburst is thought to occur because of an excess of matter piling up on the accretion disk, that is, piling up faster than it can spread through the disk by steady viscous interactions. The disk becomes unstable, expands, and excess material ends up spreading onto the white dwarf, causing an outburst. After a while the excess material dies down and normality resumes. Yoji Osaki, writing in the Publications of the Astronomical Society of Japan in 1974, was the originator of this 'disk instability' model. Professional astronomers have verified this model of a CV by studying dwarf nova outbursts in the brightest systems. The bright CV called U Geminorum (after which the class of CVs known as U Gem stars are named) is one that has had its accretion disk radius measured with some precision by professional astronomers during its outbursts. The technique uses eclipses of the bright spot on the accretion disk to measure the disk radius. Their results indicate that, as predicted by the Osaki model, the accretion disk expands dramatically to its maximum size (i.e., within a day) during an outburst and contracts slowly over the subsequent weeks and months. The mutual eclipses of such binary star systems, when in outburst (and when not) can be used to great advantage by amateur and professional observers. The reader does not have to be a genius to realize that our line of sight with respect to the orbital plane of such systems is absolutely crucial to what is observed. If the orbital plane is at right angles to our line of sight we will not see eclipses, whereas if it is parallel we will be optimally placed. It hardly needs mentioning that eclipses are the most powerful way of determining the orbital period.

Dwarf nova outburst intervals come in all sizes. The system WZ Sge, thought to be the closest CV to us at 140 light-years, outbursts every 30 years with an 8.5-magnitude range perhaps more reminiscent of a nova; at the other extreme V1159 Ori outbursts by typically 2 magnitudes every 4 days. At the end of July 2001, during a rare outburst of nearby WZ Sge, astronomers from Southampton University, Utrecht University, the University of St Andrews, and the Netherlands Space Research Organization announced they had detected spiral arm-like structures in the disk of WZ Sge during the outburst. Using phase-resolved spectroscopy with the 2.5-m Isaac Newton telescope they were able to construct Doppler tomograms to indicate what these structures actually looked like in three-dimensional (3D) form, despite the 140-light-year distance of the system.

Variable Star Nomenclature

To the beginner the variable star labeling system, whether for CVs or not, can seem highly confusing. Is there any logical system for naming these stars? Well, there is a system, but it is perhaps more historic than logical!

The vast majority of CVs, and variable stars below naked-eye visibility, have a two-letter designation, such as with VY Aqr in Aquarius, for example. Historically, well before so many variables were known, if the variable star did not already have a Greek alphabet designation, it would be given a single-letter label, starting with the capital letter R. Why R? Well, because the system was developed by Friedrich Argelander (1799–1875), who was well aware that lowercase letters and the early part of the alphabet in capital letters had already been used (as in Bayer's Uranometria of 1603). So, capital letters in the latter part of the alphabet were a safe bet, and Argelander had no idea just how many variable stars there would end up being. R–Z gives you nine letters. Surely there could not be more than nine variable stars, even in a Milky Way constellation like Cygnus?

It soon became apparent there were many more variable stars than anyone in the early nineteenth century could have dreamed of. So, to cope with more than nine variables after the letter Z, the trusty letter R was used again, but in duplicate, that is, RR to RZ, then SS to SZ all the way down to ZZ. By the time you get to ZZ you have used up 54 designations. But still, this was not enough. So finally the earlier parts of the alphabet were then roped into service, with AA to AZ, BB to BZ, and, eventually, QQ to QZ being used, and the letter J omitted throughout. This gave the designators an extra 280 star names, giving 334 in total per constellation. Sadly, even this was nowhere near enough, especially for constellations in the Milky Way, and even before the CCD era. So eventually, the slightly more boring option of just labeling the rest from V335 onward (the 335th variable in that constellation) had to be implemented.

As an indication of just how many variable stars there can be in a constellation take the example of the recent Nova in Sagittarius in 2007 discovered by Yukio Sakurai on April 14th of that year. Shortly after discovery the International Astronomical Union (IAU) designated it as V5558 Sgr. One well-known anomaly to the Argelander numbering system is the old nova designated Q Cygni. Q obviously comes before R, but the nova was discovered in 1876, well after Argelander's system had been devised in the mid-1800s. It may first have been named as such by S. C. Chandler in his 1893 variable star catalog and simply named Q as Q alphabetically follows P Cygni, a well-known variable star in that constellation. The designation has stuck, though, and Argelander's system seems to have been mysteriously ignored in this puzzling case.

As if all this was not illogical enough there is a second designation system, called the Harvard system, where a variable is allocated a six-digit number plus a sign to simply give its R.A. and Dec. position in the sky. For example, R Leonis is sometimes seen as 0942+11. This system is sometimes seen in very new discoveries. For example, new variable stars discovered by the Sloan Digital Sky Survey (SDSS) are often in this format, for example, SDSSpJ015543.40+002807.2, with the p indicating preliminary astrometry. Similar R.A. and Dec. formats are frequently seen with a 1RXS, RXS, or RX prefix (ROSAT satellite), FBS (First Byurakan Spectral Sky Survey) prefix, FSV (Faint Sky Variability Survey) prefix, EUVE (Extreme Ultraviolet Explorer) prefix, or NSV (Catalogue of New and Suspected Variables) prefix. A quick visit to the SIMBAD website (http://simbad.u-strasbg.fr/simbad/) will reveal just how many different designations astronomical objects can have, whether they are stars, or galaxies, or high-energy sources.

When a CV is in quiescence the visual magnitude of the system may be composed of similar light outputs from the white dwarf and the accretion disk. On top of this is the brightness of the hot spot that contributes a similar amount of light and often more, assuming it is not in eclipse. Of course, if the CVs orbital plane is in our line of sight the total light output can be considerably complicated by eclipses of the white dwarf by the red dwarf, eclipses of the hot spot by the red dwarf and the white dwarf, and partial eclipses of the disk by the red dwarf. The red dwarf's contribution to the total light output is usually small, as it may be considerably less luminous than the white dwarf. To all intents and purposes it is an eclipsing body.

Astronomers can learn a lot about the composition of a CV binary system (if its orbital plane is parallel to our line of sight) simply by studying its light-curve in quiescence and the effects of eclipses. Of course, the rapid rotation of such a system is a great help here, as within a matter of hours an entire orbit, or a large percentage of an orbit, is completed; thus, one long, clear night is all that is required to get a complete light-curve. If you think about it you realize that the accretion disk surrounding a white dwarf primary must lie within the Roche lobe of that star. Analyses of these systems, using eclipse data, have shown this to be correct and that the biggest accretion disks fill no more than 90 percent of the Roche lobe diameter, before gravitational (tidal) forces from the red dwarf limit further expansion.

When a CV system goes into outburst the magnitude shoots up by, typically, 5 magnitudes, although anywhere between 3 and 8 magnitudes is possible. Outbursts of CVs have been studied since the 1850s (in the case of U Geminorum), but, of course, no one would have a sound theory for the violent outburst mechanism until more than a century later. The work of Robert P. Kraft in the late 1950s, the development of high-speed photometers in the 1970s, and the powerful computer simulations of the 1990s have left us with a solid theory as to what a CV consists of and what takes place during an outburst. Once again, systems that are aligned favorably to our line of sight, enabling us to witness eclipses of the white dwarf, accretion disk, and hot spot yield the most clues. Unfortunately, even with the Hubble Space Telescope (HST), CV systems are not resolvable optically. Even the closest examples do not span 1/20th of an arcsecond, the optical resolution of HST.

During outburst the accretion disk brightens dramatically, totally outshining the energy output of the white dwarf and the hot spot. Thus, as the accretion disk diameter is far wider than the red dwarf, eclipses by the red dwarf when the CV is in outburst will not extinguish the outburst; it will still be much brighter than normal, even in favorably aligned systems.

Precisely what is happening in an accretion disk when a CV outbursts is still a matter of much debate. However, the viscosity of the material in the disk is of prime interest. In everyday situations the term viscosity is associated with materials such as lubricants and engine oil, that is, how sticky a liquid substance is. In CV accretion disks a similar stickiness seems to occur, but in such a tenuous material it is not immediately obvious why.

To explain this viscous-like behavior the theoreticians have invoked turbulence within the disk as the viscosity-impersonating mechanism. Magnetic instability has been cited as the cause of the turbulence, that is, how an ionized gas interacts with an unstable magnetic field. In this context it has proved important for the theorists to consider adjacent rings (annuli) of material in the disk of a CV and how the material in adjacent rings interacts magnetically. Of course, the material in the disk is in orbit around the white dwarf, and so an annulus of material that is closer to the primary star orbits more rapidly than an annulus that is further out. This may seem trivial on the small scale, but it is the magnetic interaction between materials in neighboring annuli, rotating at different rates, that leads to amplification and stretching of the magnetic forces. This, in turn, leads to magnetic turbulence and an effect with the characteristics of viscosity; viscosity that increases dramatically during a dwarf nova outburst.

Beyond this point the theory itself gets very sticky, but, essentially, the method by which viscosity in such a tenuous mechanism arises has been explained, in general, to the satisfaction of most astronomers, and the outburst mechanism explanation then proceeds as follows. In a quiescent, low-viscosity, relatively low-temperature (i.e., 3000 K) state the disk has a relatively low mass transfer flow rate. Then, when a faster flow of mass than it can cope with arrives from the secondary, the hydrogen becomes partly ionized (a big development in a magnetic environment), disk density increases, temperature increases, and a runaway temperature rise then occurs, to as high as 20,000 K. The disk hydrogen becomes completely ionized and a much higher inner flow of disk material, far higher than the flow supplied by the secondary red dwarf, takes place in the new high-viscosity environment. Thus, as the internal flow exceeds the supply of material, the surface density starts to fall, the hydrogen goes from ionized to partly ionized, viscosity drops, and the temperature plunges back to the pre-outburst state of around 3000 K.

Categories of Dwarf Novae

Dwarf novae (frequently just called CVs) are typically divided into three major subtypes, named after the most famous example of each category. Like most things in stellar astrophysics, though, no star is the same, and there are many intriguing cases within each subtype and many minor subtypes. The star U Geminorum provides the designation for the best-known category, primarily because it has been studied since 1855. U Gem stars are sometimes defined as being those dwarf novae that are not members of the other subtypes SU Uma (SU Ursae Majoris) or Z Cam (Z Camelopardalis). This means that U Gem stars are not prone to super-outbursts (like SU UMa types) and do not exhibit standstills in their light-curves.

U Gem itself lies in the northern hemisphere winter constellation of Gemini at R.A. 7 h 55m 5.5s and Dec. 22° 0′ 9″ and varies in magnitude between the extremes of 14.9 and 8.2 (i.e., a binocular object in outburst), with an average period of 105 days between outbursts. Sometimes the classification SS Cyg is loosely interchanged with U Gem. SS Cyg is equally well known (Figure 1.2) and lives in the northern hemisphere summer constellation of Cygnus, although

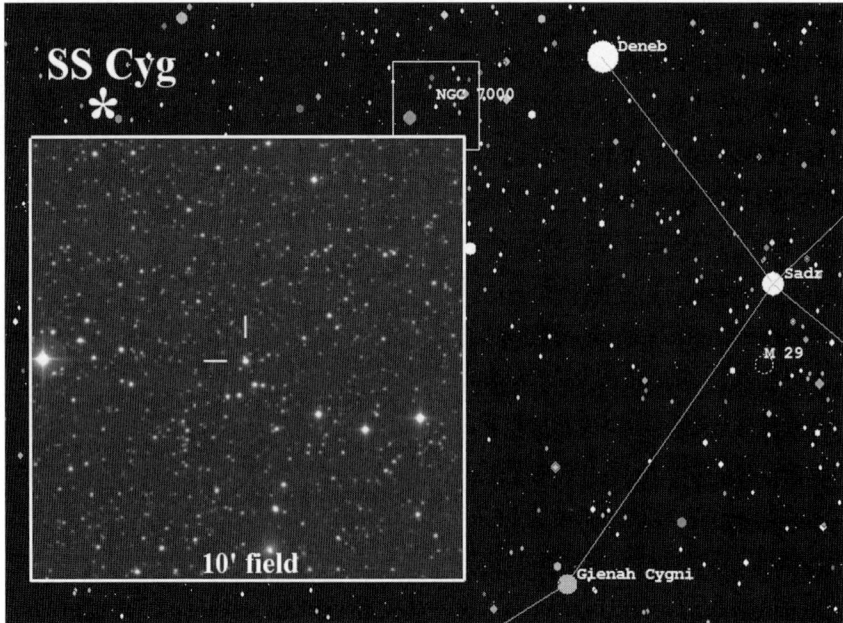

Figure 1.2. Probably the best-observed dwarf nova system, SS Cyg is well placed for most of the northern hemisphere year and is overhead in summer months, not far from the bright star Deneb and the North American Nebula. Its position is marked with an asterisk in the wide-field background shot and arrowed in the 10′ DSS field. North is up.

from high northern latitudes it is visible for virtually the whole year, not just the summer. Above latitude 47° north it is circumpolar. SS Cyg lies at a position of 21 h 42m 42.9s and Dec. +43° 35′ 10″ and varies between magnitudes 12.4 and 7.7.

The distinction between the pure U Gem and SU UMa subcategory of CV is, essentially, that the SU UMa systems have orbital periods of less than 3 hours, and the vast majority are under 2 hours. We briefly mentioned earlier the mysterious orbital period gap between 2 hours and 3 hours for CV systems. After studying some 300 CVs in detail, astronomers now think the gap, and the distribution of orbital periods, can be explained.

Roughly half of CVs have orbital periods between 78 minutes and 2 hours. Most of the remainder have periods between 3 hours and 12 hours, but there is a noticeable tail off even above 6 hours or 7 hours. The 12-hour limit has been linked to the Chandrasekhar limit of 1.4 solar masses for the white dwarf primary star and the fact that the red dwarf must be less massive than the white dwarf to avoid a catastrophic flow rate of hydrogen. Heavier stars mean larger Roche lobes and longer orbits, until the Chandrasekhar limit prevents a CV system operating with a period above 12 hours.

The discerning and knowledgeable reader will have heard of systems with longer periods, though. GK Per is, perhaps, the best example with a period of just under 48 hours. Astronomers have an answer for this apparent anomaly; they think the secondary is expanding en route to becoming a red giant star. Binary star systems with lower mass secondaries become CVs with shorter periods when magnetic braking, over hundreds of millions of years, reduces the separation between white dwarf and red dwarf until mass transfer can start. As the red dwarf shrinks from the loss of matter, and further magnetic braking takes place, we arrive at the 3-hour to 2-hour period gap. The SU UMa subtypes only live below this 3-hour boundary, and the overwhelming majority are below 2 hours.

So why is there an orbital gap at all? Theorists think it comes about because the red dwarf in such systems contracts at the 3-hour orbit point when magnetic braking ceases; it contracts due to so much material being lost to the white dwarf/accretion disk. The pressure on the red dwarf's core has reduced, in turn reducing nuclear reactions, in turn reducing the red dwarf's diameter. The red dwarf shrinks inside its Roche lobe, and the hydrogen flow from red dwarf to white dwarf turns off. Due to gravitational interactions the period continues to decrease, though and at some point, the red dwarf is, once again, close enough to the white dwarf for mass transfer to take place. This, of course, is when the orbital period has dropped to 2 hours. The magnetic braking mechanism, caused by stellar magnetic field/wind interactions is the force that has reduced the orbital period down to the 3-hour point. It seems to become ineffective at the 3-hour point, and the shut off triggers the contraction of the red dwarf.

The exact reason why this shutoff takes place is a subject for much discussion. Perhaps, because of the mass of the shrinking red dwarf, at the 3-hour point, convective forces in the star stall the internal dynamo, or maybe the 3-hour rotation rate of such a star is counterproductive to generating a magnetic field. Regardless of the reason, the period gap exists, and the CVs above it cannot be SU UMa CVs; they are U Gem CVs. U Gem CVs exist on either side of the period gap, but mainly above it. SU UMas only live below it. Below the gap the red dwarf's mass will be less than one-third of the white dwarf's mass; above the gap the red dwarf's mass will be more than one-third of the white dwarf's mass. Typically the white dwarf will have a mass of roughly two-thirds that of our Sun, compressed into an

Earth-sized sphere. Some astronomers think that the short-period (<3 hours) U Gem CVs may eventually be reclassified as SU UMa's if they are shown to feature the characteristics described next.

Superoutbursts and Superhumps

SU UMas have superoutbursts and show so-called 'superhumps' in their light output. These are the main characteristics defining SU UMa CVs. With appropriate equipment both these phenomena are recordable using amateur CCD equipment, and obtaining such data is a very worthwhile aspect of the hobby. SU UMa itself is located at 8 h 12m 27.7s +62° 36′ 24″ near to Ursa Major's boundary with Camelopardalis. It varies between magnitudes 15 and 10.8, with a typical interval of 11–17 days and superoutbursts every 153–260 days.

Superoutbursts are, not surprisingly, rather brighter than the normal outbursts of an SU UMa CV, and they last for a longer duration. Perhaps more surprising is that they tend to be more regular. The superhumps are a more subtle phenomenon, only revealed by accurate photometry. These are smaller outbursts on top of the existing superoutburst, with a period fractionally longer than the known orbital period of the CV. The cause of superhumps is thought to be related to the accretion disk in SU UMa stars becoming elliptical during a superoutburst, because of the disk's interaction with the secondary star. When the red dwarf's orbit takes it past the long axis of the ellipse, say, every 100 minutes, during a superoutburst, a superhump will occur. If the elliptical disk is precessing slowly around the white dwarf in a period of days, the encounter with the elliptical long axis will occur slightly later each orbit, and the superhump will occur a few minutes later each orbit. In fact, this is exactly what is observed: the superhump period interval is a few minutes longer than the known orbital period.

Why should the red dwarf's encounter with the elliptical disk cause a superhump in the light-curve at all? Well, powerful computer simulations indicate that in the inner accretion disk the disk particle orbits remain circular, while the outer accretion disk orbits have become elliptical. As the secondary star passes the tidal bulge in the disk, orbits start to seriously intersect and particles collide, with a brightening developing mainly at the disk edge facing the secondary star.

Moving back now to superoutbursts (as opposed to superhumps; it is easy to confuse the two terms), the time interval between SU UMa superoutbursts is referred to as the supercycle and is typically several hundred days in length. However, there are considerable variations of this SU UMa supercycle period, and the extreme examples of these groups are known as WZ Sge stars (long supercycles of decades) and ER UMa stars (short supercycles of 20–50 days). Predictably the major differences are down to the actual mass transfer rates from red dwarf to white dwarf. WZ Sge can be found at 20 h 07m 36.41s +17° 42′ 15.4″ and has a superoutburst every 33 years. ER UMa lives at 09 h 47m 11.93s +51° 54′ 09.0″ and its supercycle period is 43 days.

Nova-likes, Z Cams, and Others

Nova-like variables and Z Cam stars are two other varieties of CV (novae themselves are dealt with in Chapter 2). The star UX UMa is often cited as the defining nova-like variable. UX UMa lives not far from Alkaid, the tail star of the Great Bear, at 13 h 36m

41s, 51° 54′ 49″ and varies between magnitudes 14.2 and 12.6. There are considerable differences between novae and nova-like variables, but the term nova-like variable implies that the CV's accretion disk is, technically, permanently in outburst, that is, in the hot, viscous mode described earlier. This permanent outburst state implies high mass transfer volumes from red dwarf to accretion disk, which implies that a nova-like system lies above the period gap. In fact, not all lie above the gap! Trying to conveniently box every CV into a well-behaved category simply does not work. SW Sex stars (named after a star in the southern constellation Sextans) are thought to have the highest mass transfer rates of any subtype of nova-like CVs. They have some strange properties, such as not exhibiting Doppler-shifted double-peaked emission lines from opposite ends of their disks. Only one emission line is seen, even in edge-on disk systems. Astronomers think this may be due to an accretion stream flowing over the accretion disk or a flared accretion disk.

Some nova-like variables show permanent superhumps in their light-curves. VY Sculptoris (VY Scl) stars are nova-like variables with, much of the time, a hot disk and a high mass transfer rate. They have been called 'anti-dwarf novae,' a rather misleading term. VY Scl stars can drop by several magnitudes into a low state, where mass transfer appears to completely cease. Some astronomers have speculated that this property may be related to a hotter than normal white dwarf primary bathing the inner accretion disk so it has a higher accretion rate and less mass when the low state commences. More evidence is required, though, and as always, both amateur and professional results add to our knowledge year after year.

Z Cam stars are thought to have a mass transfer rate from red dwarf to accretion disk just below that of nova-like variables, and above that of U Gem stars; in other words, the accretion disk hovers on the limit between hot outburst state and the cooler quiescent state. Z Cam itself lies a hair's breadth above the Ursa Major/ Camelopardalis border at 8 h 25m 13.3s +73° 6′ 40.0″. The light-curve of Z Cam shows it sitting at magnitude 11.5 for long periods, sometimes more than 18 months, but it also shows manic outburst periods when the star swings between magnitudes 10 and 13.5 with a period frequency of weeks, or up to 2 years in the most extreme outburst periods. The long periods of calm in Z Cam stars' light-curves are called standstills. In 2003 NASA's Galaxy Evolution Explorer, imaging in the near and far ultraviolet region of the spectrum, detected a massive shell around Z Cam, hinting that maybe 2000 years ago it had experienced an eruption more akin to a full-blown classical nova outburst; this came as a surprise to many astronomers. Following the detection of this shell, Goran H. I. Johansson of the University of Lund, Sweden, suggested that a 'guest star' detected by the Chinese in 77 B.C. was actually Z Cam in full nova outburst. He points out that at a distance of 530 light-years the nova may have been as bright as Venus and be the oldest known recorded nova outburst (the next oldest being seventeenth-century events).

We have already mentioned that CVs have orbital periods between 78 minutes and 12 hours, but, in fact, that is not really true! In normal CV systems the 78-minute barrier corresponds to a CV where the red dwarf's mass has become so low that it more resembles a white dwarf. Powerful computer simulations of CVs show that these systems cannot evolve to shorter periods, but only to longer ones. However, in the class of CVs known as AM Canes Venaticorum systems, or AM CVn's, the secondary is mainly composed of helium, not hydrogen. The same applies to the accretion disk. Because helium is denser than hydrogen, the Roche lobe contact point is closer. It may seem like a statement from a science fiction film, but the two stars in these systems orbit each other in periods as short as 15 minutes!

These systems exhibit permanent superhumps. One of the best-known examples is CR Boo, whose magnitude ranges between 13.5 and 17.5 and the outbursts are frequent, typically every day or two. The orbital period is a mere 24 minutes and 31 seconds. Should CR Boo be considered the most prolific dwarf nova in the sky?

A Period–Luminosity Link

In 2003 astronomers Joni J. Johnson, Thomas E. Harrison, and colleagues at New Mexico State University used the Fine Guidance Sensors of the HST to pin down the distances to some well-known dwarf novae from their parallactic shifts. By using Earth's orbit around the Sun to get the star's positions to shift against more remote stars, astronomers can estimate their distances. The positional shifts are tiny, ±1 arcsecond for every 3.26 light-years of distance; hence, such a distance is known as a parsec. Nevertheless, HST can measure such parallaxes with errors of only ±0.5 milli-arcseconds, 100× smaller than the telescope's actual optical resolution. This enables objects at distances of even 1000 light-years to be estimated to an accuracy of roughly ±15%.

The results confirmed the distances to U Gem and SS Cyg mentioned earlier, and that WZ Sge is the closest known CV to us at 140 light-years (and possibly the closest known system with an accretion disk). But it also revealed that many of the secondary stars in the systems studied were rather brighter than originally thought, making them further away. The New Mexico team also found a good correlation between a dwarf nova's orbital period and its outburst luminosity: the longer the orbital period the brighter the outburst luminosity, when factors such as orbital inclination were taken into account. Thus, WZ Sge itself, with an orbital period of 1 hour 22 minutes, has a relatively dim absolute outburst magnitude of 5.5, whereas RU Peg, with an orbital period of almost 9 hours, has an absolute outburst magnitude of 1.2, or 50× brighter. If this relationship holds then future estimates of dwarf nova distances may be possible simply by using the outburst magnitude, orbital period, and orbital inclination.

Magnetic CVs

In CV systems where the white dwarf has a substantial multimillion Gauss magnetic field, ionized material from the red dwarf heads straight for that star's magnetic poles (Figure 1.3), that is, in a not too dissimilar way to that by which charged particles head for Earth's poles and create the aurora. However, in the CV case the magnetic field strength is colossal, namely tens of millions of times greater than that of Earth. An accretion disk cannot form in such a powerful magnetic field, but material accretes vertically downward onto the white dwarf, leading to intense radiation in the extreme ultraviolet and X-ray frequencies. In the so-called polars, or AM Herculis stars, the magnetic field has linked the red dwarf and white dwarf together such that they rotate like a fixed dumbbell, like Pluto and Charon.

Polars emit light that is linearly and circularly polarized. Intermediate polars have lower magnetic fields, and the white dwarf spins faster than the orbital period. In these cases there can be an accretion disk, but it is disrupted by the magnetic field, or there can be an accretion disk and an accretion stream onto the white dwarf. In fact, various permutations are possible, and understanding each

Figure 1.3. In CV systems where the white dwarf has a colossal magnetic field, ionized material from the red dwarf heads straight for that star's magnetic poles rather than to an accretion disk. Graphic by the author.

case is a major challenge for computer simulations. Intermediate polars with the fastest spin speeds are known as DQ Herculis (DQ Her) stars. AM Her itself is situated at 18 h 16m 13.6s and +49° 51′ 56″ and varies between magnitudes 15.7 and 12.3. The star XY Arietis is the only fully eclipsing intermediate polar star known, but, tragically, it is hidden behind a molecular cloud called Lynds 1457 that probably attenuates its light by 11 or 12 magnitudes. Nevertheless, it has been studied by the Chandra orbiting X-ray telescope.

The Hamburg Quasar survey (Quasars are covered in Chapter 5) picked up not only Quasars (i.e., active galaxies) but a large number of magnetic CVs too. A portion of these have been selected by Dr Boris Gaensicke of the U.K.'s Warwick University as suitable for monitoring by amateur astronomers and are listed in Table 1.1. To acquire more information on these CVs, detected by the Hamburg

Table 1.1. The BAAVSS Long Term Polar Monitoring Programme has been set up to monitor a selection of magnetic AM Her/Polar stars that are in need of further investigation

Star	R.A. (2000)	Dec. (2000)	Range	Orbital period
SDSSJ015543+002807	01 h 55m 43.44s	+00° 28′ 06.4″	14.7–17.6	01 h 27m
Al Tri	02 h 03m 48.60s	+29° 59′ 26.0″	15.5–18.0	04 h 36m
V1309 Ori	05 h 15m 41.41s	+01° 04′ 40.4″	15.2–17.3	07 h 59m
BY Cam	05 h 42m 48.77s	+60° 51′ 31.5″	14.6–17.5	03 h 21m

(Continued)

Table 1.1. The BAAVSS Long Term Polar Monitoring Programme has been set up to monitor a selection of magnetic AM Her/Polar stars that are in need of further investigation (*Continued*)

Star	R.A. (2000)	Dec. (2000)	Range	Orbital period
GG Leo	10 h 15m 34.67s	+09° 04′ 42.0″	16.5–18.8	01 h 20m
AN UMa	11 h 04m 25.67s	+45° 03′ 14.0″	13.8–20.2	01 h 55m
ST LMi	11 h 05m 39.76s	+26° 06′ 28.7″	15.0–17.2	01 h 55m
AR UMa	11 h 15m 44.56s	+42° 58′ 22.4″	13.3–16.5	01 h 56m
DP Leo	11 h 17m 15.93s	+17° 57′ 42.0″	17.5–19.5	01 h 30m
EU UMa	11 h 49m 55.71s	+28°45′ 07.6″	16.5–faint	01 h 30m
MR Ser	15 h 52m 47.18s	+18° 56′ 29.2″	14.9–17.0	01 h 53m
AP CrB	15 h 54m 12.34s	+27° 21′ 52.4″	16.8–faint	02 h 32m
1RXSJ161008+035222	16 h 10m 07.51s	+03° 52′ 33.0″	15.9–faint	03 h 10m
V2301 Oph	18 h 00m 35.58s	+08° 10′ 13.6″	16.1–21.0	01 h 53m
V884 Her	18 h 02m 06.52s	+18° 04′ 44.9″	14.5 V–faint	03 h 43m
AM Her	18 h 16m 13.33s	+49°52′ 04.3″	12.0–15.5	03 h 06m
QQ Vul	20 h 05m 41.91s	+23° 39′ 58.9″	14.5–15.5	N/A

The stars above, with the exception of AM Her (included as it is the definitive star for this category of object) should be observed on a nightly basis both visually and with CCDs. Any changes should be reported. The program is supported by Dr Boris Gaensicke of Warwick University, U.K., whose article on polars appeared in the September 2006 issue of the BAAVSS Circular (No. 129). Dr Gaensicke notes that: "Al Tri and V1309 Ori are two very long period polars. BY Cam contains a slightly asynchronously rotating white dwarf. AR UMa, AP CrB, and V884 Her are Polars with very high magnetic fields. SDSS015543 is deeply eclipsing."

Quasar Survey, see the Warwick University webpages at http://deneb.astro. warwick.ac.uk/phsdaj/HQS_Public/HQS_Public.html.

Symbiotic Stars

Finally, at this point, it is worth mentioning the category of binary stars known as symbiotic stars. Although these are different to CVs, there are major similarities in the binary components, as they consist of a white dwarf and a larger red star, but in this case the red star is losing its mass not through the standard Roche Lobe/accretion disk mechanism seen in CVs but via a spherically symmetric wind. As with specific types of CVs it is thought that symbiotic systems with white dwarfs close to the 1.4-solar-mass limit might evolve to become Type Ia supernovae. Symbiotic stars under amateur scrutiny include Z And, EG And, V1413 Aql, CM Aql, V1329 Cyg BF Cyg, CI Cyg, YY Her, and BX Mon. Most symbiotic stars spend long periods of time in quiescence interrupted by brief activity or a few months in a high state.

How to Observe CVs

For the beginner, learning to make variable star magnitude estimates, and even finding CVs in the sky, requires patience and practice over months and years. Absolutely crucial to this process is mixing with a like-minded group of people and asking for help from the veterans of the hobby when something does not make sense. I cannot stress this aspect enough. Climbing the mountain of useful CV observing will be almost impossible unless you have contacts to bounce questions off and an e-mail alert service telling you what is the hottest outburst around. The Internet is also a vital source of information on star charts and photometric advice (see Chapter 7 for more on advanced photometry).

Online CV Information and Star Charts

Fortunately, the Internet has meant that acquiring CV information, and asking for help, has never been easier. Internet astronomy forums are on hand everywhere, and experts can advise you by e-mail within minutes or hours of a query. The two biggest amateur organizations involved in variable star research are the American Association of Variable Star Observers (AAVSO) and the British Astronomical Association's (BAA's) variable star section (VSS). In addition, the U.K.-based group The Astronomer publishes variable stars results every month, and the Center for Backyard Astrophysics (CBA) is very active in CV photometry. The newer CVnet along with the BAA's VSS-alert system, both on Yahoo, are vital sources of information, too. The most relevant webpages for these organizations are listed below:

BAA VSS: www.britastro.org/vss/

AAVSO: www.aavso.org/

The Astronomer: www.theastronomer.org/

The Center for Backyard Astrophysics: http://cba.phys.columbia.edu/

CVnet is probably the first place to go for the beginner who wants to mingle in cyberspace with the hardcore CV enthusiasts (amateurs and professionals). The main information page is at Mike Simonsen's site, which http://cvnet.aavso.org will take you to, and we will refer to Mike again shortly. But there are also useful CVnet discussion, outburst, and circular sites on the search engine Yahoo that can be browsed online or used to generate individual or summarized e-mails to your mailbox. To subscribe to these free discussion groups and outburst and circular facilities go to the following sites:

http://tech.groups.yahoo.com/group/cvnet-discussion/

http://tech.groups.yahoo.com/group/cvnet-outburst/

http://tech.groups.yahoo.com/group/cvnet-circular/

The BAA VSS alert group covers all unusual variable star activity, but CVs get the bulk of coverage and much (but not all) of the dialog is from British observers:

http://tech.groups.yahoo.com/group/baavss-alert/

Of course, for every variable star observation you need a star chart with which to identify the object and compare its brightness with photometrically measured stars. It would be impractical to include a chart for every object covered in this book, and charts sometimes alter as the years go by, that is, when better sequences become available. The Internet is a wonderful source of up-to-date star charts. The AAVSO site at www.aavso.org/observing/charts/ has been a vital source of charts for many years. You simply type in the variable star's name, and, with luck, a choice of charts for various field sizes and a direct or mirror-flipped view is available. At the time of writing the AAVSO has a new facility that will supersede the old webpage and is called the VSP or Variable Star Plotter. This service enables any area of the sky to be called up with comparison stars for any R.A. and Dec. or any known star name. Essentially, the system draws you a star chart to suit your preferences. This system was, in 2008, at the Beta test level and will probably be the main source of variable star charts in the coming years. The VSP is located at www.aavso.org/observing/charts/vsp/.

As an aside, if you are toying with investing in some flashy new 'Go To' equipment to study CVs, bear in mind that some telescope hand controllers have far larger databases than others, and variable stars are often given a lower priority than deep-sky objects. Of course, if you are using a laptop running software such as Bisque's The Sky or Project Pluto's Guide 8.0 to slew the telescope, this may well be academic, but the newest objects may not appear even on those massive databases, except as an obscure star number. Sometimes it can be easier to use your Go To hand paddle to slew to a nearby deep-sky feature or other variable star first, and then crawl from there to the new object.

When submitting observations of any variable stars to your preferred organization you will need to know the contact name and e-mail address and the preferred format. The precise method of quoting variable star estimates is discussed in Chapter 7. Members of the AAVSO can submit their observations via the slick Webobs interface online. This can be found at www.aavso.org/bluegold/index.php and is known as 'blue and gold,' a reference to the level of your AAVSO membership. You do not need to be an AAVSO member to submit observations, but you do need to register with a username and password. The 'bluegold' area is the password-protected area of the AAVSO website where you can submit an observation or update your contact information, access robotic telescopes, and more. You do not need to be a paying member of the AAVSO to enter observations, but if you are a member you will have access to more features.

The First CV Observation

The first dwarf nova to be discovered and observed was U Geminorum, on December 15, 1855. It was discovered by J. R. Hind, a prolific English astronomer who, in 1844, left Greenwich Observatory and took over from William Rutter Dawes as director of George Bishop's private observatory, situated in London's Regent's Park. These days, a worse place to house an observatory is hard to imagine! The observatory featured a high-quality 7-inch refractor by Dollond, mounted on a massive English mounting and complete with the obligatory observing chair/couch popular at the time. Hind reported his discovery of U Gem, in the Monthly Notices of the RAS 16, 56 (1856) as follows:

> On the evening of December 15th, 1855, I remarked in R. A. (1856) 7 h 46m 33s.65, N.P.D. 67° 37′ 17″.1, an object shining as a star of the ninth magnitude, with a very blue planetary light, which I have never seen before during the five years that my attention has been directed to this quarter of the heavens. On the next fine night, Dec. 18, it was certainly fainter than on the 15th by half a magnitude or more. Since that date I have not had an opportunity of examining it till last evening, January 10th, when its brightness was not greater than that of stars of the twelfth magnitude. It is evidently a variable star of a very interesting description, inasmuch as the minimum

brightness appears to extend over a great part of the whole period, contrary to what happens with Algol and S Cancri.

The position given above was deduced by micrometrical comparisons with the principal component of the double star Σ1158. The variable precedes 1m 26s·53, and is N. 7′ 30″·8.

Mr. Bishop's Observatory, 1856, January 11.

In the twenty-first century U Geminorum is still a great starting place for observing dwarf novae. As Hind witnessed more than 150 years ago, U Gem can outburst to 9th magnitude, but its quiescent state resides at 14th magnitude. The orbital inclination of 70° from our perspective is good enough for eclipses of the disk to be visible (90° being a perfect eclipse scenario for eclipses of disk and white dwarf), and during disk eclipses by the red dwarf the system can fade to magnitude 15. It takes 4 hours and 15 minutes for the two components of U Gem to orbit one another, and astronomers have a good idea of what the component stars are like. The white dwarf in U Gem is thought to have a mass of 1.2 solar masses, and the red dwarf mass is estimated at 0.4 solar masses. Their respective orbital velocities are around 100 km per second and 300 km per second.

According to HST measurements, U Gem lies at a distance of about 96 parsecs, give or take 5 parsecs, and as a parsec equals 3.26 light-years, this translates to the dwarf nova lying at almost 315 light-years from Earth. A simple inverse square law calculation tells us that the U Gem system is 100× fainter than our own Sun in quiescence, but can outburst to double the brightness. Our own Sun has an absolute magnitude (brightness seen from 10 parsecs) of 4.83. As mentioned earlier, U Gem lies in the northern hemisphere's winter constellation of Gemini at R.A. 7 h 55m 5.5s and Dec. 22° 0′ 9″; it outbursts, on average, every 120 days or so, but the extreme durations between outbursts range from 71 days to 223 days. The outbursts come in two types. In the first type it stays above magnitude 11 for 3–6 days, but in the wider case it stays there for 10–34 days. The latter outbursts are slightly brighter. With a large-aperture amateur telescope, or a more modest telescope equipped with a CCD, the eclipses can be observed every long, clear winter's night in the northern hemisphere.

However, the beginner may prefer a slightly brighter target and one visible in the northern hemisphere's summer skies. In this case the star SS Cygni is the one to watch (refer again to Figure 1.2). With a minimum brightness of magnitude 12.2, SS Cyg is easy to observe visually even in a modest-aperture telescope, all the way through its activity cycle. Despite being brighter both at maximum and at minimum than U Gem, SS Cyg was the second dwarf nova to be discovered. It was found in the early days of photographic patrolling, by Louisa Wells at Harvard Observatory. SS Cyg lies at 21 h 42m 42.9s and Dec 43° 35′ 10″, roughly 10 degrees east of brilliant Deneb and not far from the Cygnus/Lacerta border.

Since its discovery, every outburst of SS Cyg (and there have been more than 800) has been monitored, which is mainly due to its high northerly declination, making it circumpolar from North America and northern Europe. SS Cyg is in quiescence for three-quarters of its life. When it outbursts it reaches maximum brightness (magnitude 8.3) in a day, and the light-curve alternates between wide and narrow outbursts. Outbursts occur every 4–10 weeks, typically lasting 1–2 weeks. According to measurements by the HST, SS Cyg lies at a distance of roughly 166 parsecs (or 540 light-years) from Earth, which is quite a bit further than previously thought from its brightness. Indeed, if this figure is correct and it is 70 percent further away than U Gem and at least a magnitude brighter in outburst, the SS Cyg system must be significantly more luminous. Taking the 540 light-year figure as being correct implies that the SS Cyg system has an absolute magnitude of about 2 in outburst and 6 in quiescence. Either way, it is the easiest dwarf nova to observe in outburst from the northern hemisphere. Up to September 2007 we could also have got away with stating it was the brightest, but on September 5th/6th a newly discovered CV, now designated V455 And, rose spectacularly to magnitude 8.2. We will have more to say about this object later.

Hundreds of CVs

With more than 900 CVs in the sky (and that is without counting novae), the beginner might be fairly confused as to what objects to specialize in. There is a list of almost 200 in Table 1.2, and those marked with the letter 'R' in the right-hand

Table 1.2. An alphabetical listing of 191 of the brightest or most easily observable cataclysmic variables, comprising mainly dwarf novae with a few novae and recurrent novae of interest thrown in

Star	RA (2000)	Dec (2000)	V mag	Type	Orbit	Inc	R
AR And	01 h 45m 03.28s	+37° 56' 32.7"	11.0–17.6	UG	03 h 55m		
DX And	23 h 29m 46.68s	+43° 45' 04.1"	10.9–16.4	UG	10 h 34m		
FO And	01 h 15m 32.14s	+37° 37' 35.6"	13.5–17.5	UGSU	01 h 43m		
IW And	01 h 01m 08.91s	+43° 23' 25.8"	14.2–17.4	UGZ	N/A		
LL And	00 h 41m 51.46s	+26° 37' 21.3"	13.0–20.0	UGSU	01 h 19m		R
LS And	00 h 32m 10.21s	+41° 58' 10.4"	11.7–20.5	NA	N/A		R
LX And	02 h 19m 44.09s	+40°27' 22.3"	13.5–16.4	UG	03 h 37m		
PQ And	02 h 29m 29.54s	+40° 02' 40.2"	10.1–19	UGWZ	01 h 20m		R
RX And	01 h 04m 35.54s	+41° 17' 57.8"	10.9–12.6	UGZ	05 h 02m		
V402 And	00 h 11m 07.30s	+30° 32' 36.0"	15.4–20.3	UGSU	1 h 29m		R
V455 And	23 h 34m 01.55s	+39° 21' 42.9"	8.2–16.4	UGWZ	01 h 21m	75°	R
CI Aql	18 h 52m 03.59	–01° 28' 39.3"	8.8–15.6	NR	14 h 50m	75°	R
FO Aql	19 h 16m 38.17s	+00° 07' 37.0"	13.6–17.5	UGSS	N/A		
KX Aql	19 h 33m 53.64s	+14° 17' 46.1"	12.5–18.4	UGSU	01 h 27m		R
V725 Aql	19 h 56m 45.00s	+10° 49' 33.3"	13.7–17.3	UGSU	02 h 16m		R
HU Aqr	21 h 07m 58.22s	–05° 17' 40.1"	15.3–19.8	NL/AM	02 h 05m	80°	
VY Aqr	21 h 12m 09.26s	–08° 49' 37.1"	8.0–17.5	UGWZ	01 h 31m		R
SV Ari	03 h 25m 03.34s	+19°49' 52.9"	12.0–22.0	N/NR	N/A		R
XY Ari	02 h 56m 08.10s	+19° 26' 34.0"	Invisible!	IP/DQ	06 h 04m	84°	
SS Aur	06 h 13m 22.43s	+47° 44' 25.3"	10.5–14.5	UG	04 h 23m		
1502+09 Boo	15 h 04m 41.78s	+08° 47' 54.1"	N/A–18.6	UG	N/A		R
CR Boo	13 h 48m 55.22s	+07° 57' 35.7"	13.0–17.5	AM CVn	00 h 25m		
UZ Boo	14 h 44m 01.21s	+22° 00' 54.8"	11.5–20.4	UGSU/WZ	N/A		R
Z Cam	08 h 25m 13.18s	+73° 06' 39.2"	10.5–14.8	UGZ	06 h 57m		
AF Cam	03 h 32m 15.48s	+58° 47' 22.5"	13.4–17.3	UG	07 h 47m		

(Continued)

Table 1.2. An alphabetical listing of 191 of the brightest or most easily observable cataclysmic variables, comprising mainly dwarf novae with a few novae and recurrent novae of interest thrown in (*Continued*)

Star	RA (2000)	Dec (2000)	V mag	Type	Orbit	Inc	R
HT Cam	07 h 57m 01.37s	+63° 06' 01.8''	11.9–18.2	UG/DQ	01 h 26m		
LU Cam	05 h 58m 17.88s	+67° 53' 46.4''	14.0–17.0	UG	03 h 36m		
LS Cam	05 h 57m 23.93s	+72° 41' 52.6''	16.7–19.5	SW Sex	03 h 26m		
OY Car	10 h 06m 22.10s	–70° 14' 04.7''	12.4–17.3	UGSU	01 h 31m	83°	
AM Cas	02 h 26m 23.45s	+71° 18' 31.4''	12.3–15.2	UGZ	03 h 58m		
DK Cas	00 h 18m 07.83s	+57° 26' 07.1''	14.8–18.6	UGSS	N/A		R
GX Cas	00 h 49m 01.55s	+56° 52' 44.1''	13.3–18.5	UGSU	02 h 08m		
HT Cas	01 h 10m 13.12s	+60° 04' 35.7''	10.8–18.4	UGSU	01 h 46m	81°	
KU Cas	01 h 31m 02.37s	+57° 54' 13.4''	13.3–19.7	UGSS	N/A		
V452 Cas	00 h 52m 18.06s	+53° 51' 50.1''	14.5–18.7	UGSU	N/A		
V630 Cas	23 h 48m 51.91s	+51° 27' 39.4''	12.3–17.1	UG/WZ	61 h 32m		R
V635 Cas	01 h 18m 32.0s	+63° 44' 33''	14.2–15.7	HMXB	24.3 days		
BV Cen	13 h 31m 19.48s	–54° 58' 33.5''	10.5–13.3	UG	14 h 39m		
V803 Cen	13 h 23m 44.54s	–41°44' 29.7''	13.2–17.4	AM CVn	00 h 27m		
CG Cep	23 h 10m 26.09s	+66° 33' 31.9''	14.5–17.2	UGSS	N/A		
FX Cep	21 h 03m 06.61s	+66° 10' 32.8''	15.0–17.5	UG	N/A		
FL Cet	01 h 55m 43.48s	+00° 28' 06.7''	15.5–18.0	NL	N/A		R
WX Cet	01 h 17m 04.18s	–17° 56' 22.6''	9.5–18.5	UGSU/WZ	01 h 24m		R
Z Cha	08 h 07m 27.77s	–76° 32' 00.8''	11.9–17.2	UGSU	01 h 47m	82°	
CG CMa	07 h 04m 05.23s	–23° 45' 34.3''	13.7–<20	UGWZ/SU	N/A		R
SV Cmi	07 h 31m 08.40s	+05° 58' 48.3''	13.0–16.9	UGZ	03 h 45m		
AT Cnc	08 h 28m 36.92s	+25° 20' 03.0''	12.7–16.2	UGZ	04 h 50m		
EG Cnc	08 h 43m 03.99s	+27° 51' 49.7''	11.4–17.7	UGWZ	01 h 25m		R
GY Cnc	09 h 09m 50.54s	+18° 49' 47.5''	12.5–17.8	UG	04 h 13m	77°	
SY Cnc	09 h 01m 03.34s	+17° 53' 56.1''	11.1–14.5	UGZ	09 h 07m		
YZ Cnc	08 h 10m 56.65s	+28° 08' 33.3''	10.5–15.5	UGSU	02 h 05m		
AL Com	12 h 32m 25.80s	+14° 20' 42.3''	12.7–20.8	UGWZ	01 h 22m		R
GO Com	12 h 56m 37.09s	+26° 36' 43.5''	13.1–20	UGSU	01 h 35m		
IR Com	12 h 39m 32.03s	+21° 08' 06.9''	13.5–18.7	UGSU	02 h 05m	80°	R
T CrB	15 h 59m 30.16s	+25° 55' 12.6''	2.0–11.3	NR	228 days		R
VW CrB	16 h 00m 03.70s	+33° 11' 14.1''	14.0–<18	UGSU	01 h 18m		
AM CVn	12 h 34m 54.63s	+37° 37' 44.2''	13.7–14.2	AM CVn	00 h 17m		
EM Cyg	19 h 38m 40.11s	+30° 30' 28.3''	12.5–14.5	UGZ	06 h 59m	67°	
EY Cyg	19 h 54m 36.74s	+32° 21' 55.1''	11.4–15.5	UGSS	11 h 01m		
SS Cyg	21 h 42m 42.80s	+43° 35' 09.9''	8.0–12.1	UGSS	06 h 36m		
V337 Cyg	19 h 59m 53.0s	+39° 14' 02.0''	14.4–<16.5	UGSU	N/A		R
V404 Cyg	20 h 24m 03.78s	+33° 52' 03.2''	11.0–20.5	HMXB	6.5 days		R
V503 Cyg	20 h 27m 17.40s	+43° 41' 22.7''	13.4–17.6	UGSU	01 h 52m		
V516 Cyg	20 h 47m 09.78	+41° 55' 26.5''	13.8–16.8	UGSS	N/A		
V751 Cyg	20 h 52m 12.78s	+44° 19' 26.1''	13.6–16.3	UG/VY	3 h 28m		
V795 Cyg	19 h 34m 34.09s	+31° 32' 11.9''	13.4–17.9	UGSS	N/A		
V1060 Cyg	21 h 07m 42.18s	+37° 14' 08.8''	13.5–18.0	UGSS	N/A		
V1113 Cyg	19 h 22m 42.07s	+52° 43' 59.1''	14.0–19.3	UGSU	N/A		

(*Continued*)

Table 1.2. An alphabetical listing of 191 of the brightest or most easily observable cataclysmic variables, comprising mainly dwarf novae with a few novae and recurrent novae of interest thrown in (*Continued*)

Star	RA (2000)	Dec (2000)	V mag	Type	Orbit	Inc	R
V1251 Cyg	21 h 40m 54.4s	+48° 39′ 43.3″	12.5–<16.0	UGSU	N/A		R
V1316 Cyg	20 h 12m 13.62s	+42° 45′ 51.5″	14.1–17.6	UGSU	N/A		R
V1363 Cyg	20 h 06m 11.53s	+33° 42′ 37.7″	13.0–<17.6	UGZ/VY	N/A		R
V1454 Cyg	19 h 53m 38.45s	+35° 21′ 45.8″	13.9–20.5	UGSS	N/A		R
V2176 Cyg	19 h 27m 11.72s	+54° 17′ 50.9″	13.3–19.9	UGSU/WZ	N/A		R
AB Dra	19 h 49m 06.51s	+77° 44′ 22.9″	12.3–15.8	UGZ	03 h 39m		
CG Dra	19 h 07m 32.78s	+52° 58′ 28.9″	15.0–17.5	UG	~5 h		
DO Dra	11 h 43m 38.50s	+71° 41′ 20.7″	10.2–16.7	IP	03 h 58m		
DV Dra	18 h 17m 23.10s	+50° 48′ 18.1″	15.0–<21	UGWZ	N/A		R
ES Dra	15 h 25m 31.79s	+62° 01′ 00.2″	13.9–16.3	UGSU	~3 h?		
EX Dra	18 h 04m 14.23s	+67° 54′ 12.4″	12.5–17.2	UG	05 h 02m	86°	
IX Dra	18 h 12m 31.48s	+67° 04′ 46.0″	15.0–18.5	UGSU	01 h 36m		
KV Dra	14 h 50m 38.31s	+64° 03′ 28.4″	11.8–17.1	UGSU	01 h 25m		R
CP Eri	03 h 10m 32.75s	−09° 45′ 06.1″	16.5–19.7	AM CVn	00 h 28m		
XZ Eri	04 h 11m 25.78s	−15° 23′ 23.6″	14.6–18.7	UGSU	01 h 28m	75°	
U Gem	07 h 55m 05.23s	+22° 00′ 05.1″	9.1–15.2	UGSS	04 h 15m	70°	
CI Gem	06 h 30m 05.86s	+22° 18′ 50.7″	14.7–21.0	UGSU/SS	N/A		R
IR Gem	06 h 47m 34.69s	+28° 06′ 22.4″	11.2–17.0	UGSU	01 h 38m		
AH Her	16 h 44m 10.01s	+25° 15′ 02.0″	11.3–14.7	UGZ	06 h 12m		
AM Her	18 h 16m 13.33	+49° 52′ 04.3″	12.0–15.5	AM	03 h 06m	60°	
CH Her	18 h 34m 46.35s	+24° 48′ 02.7″	13.5–17.0	UG	N/A		
DQ Her	18 h 07m 30.26s	+45° 51′ 32.1″	1.3–14.6	NA	04 h 39m	87°	
V478 Her	17 h 21m 05.60s	+23° 39′ 36.8″	15.5–17.1	UGSU	N/A		R
V589 Her	16 h 22m 07.15s	+19° 22′ 36.7″	14.1–<18.0	UGSU	N/A		
V592 Her	16 h 30m 56.45s	+21° 16′ 58.4″	12.3–<21.0	UGSU/WZ	N/A		R
EX Hya	12 h 52m 24.22s	−29° 14′ 56.0″	9.6–<14.0	UGSU/DQ	01 h 38m	78°	
AY Lac	22 h 22m 22.1s	+50° 23′ 40.0″	14.0–<21.0	UGWZ/NR	N/A		R
T Leo	11 h 38m 26.83s	+03° 22′ 07.1″	10.0–15.9	UGSU	01 h 25m		
U Leo	10 h 24m 03.30s	+14° 00′ 11.0″	10.5–<15	N	3 h/6 h?		R
X Leo	09 h 51m 01.46s	+11° 52′ 31.3″	12.4–16.5	UG	03 h 57m		
RZ Leo	11 h 37m 22.25s	+01° 48′ 58.6″	10.5–19.2	UGSU/WZ	01 h 49m		R
GW Lib	15 h 19m 55.43s	−25° 00′ 25.3″	9.0–18.5	UGWZ	01 h 17m		
HP Lib	15 h 35m 53.10s	−14° 13′ 12.1″	13.6–13.7	AM CVn	00 h 18m		
RU Lmi	10 h 02m 07.46s	+33° 51′ 00.3″	13.8–19.5	UG	06 h 01m		
RZ Lmi	09 h 51m 48.93s	+34° 07′ 23.9″	14.2–16.8	UGSU	01 h 24m		
SS Lmi	10 h 34m 05.43s	+31° 08′ 08.3″	15.0–21.6	UG/N	N/A		R
SX Lmi	10 h 54m 30.43s	+30° 06′ 10.2″	13.0–17.4	UGSU	01 h 37m		
AY Lyr	18 h 44m 26.69s	+37° 59′ 52.0″	12.3–18.0	UGSU	N/A		
CY Lyr	18 h 52m 41.38s	+26° 45′ 31.5″	13.2–17.0	UG	03 h 49m		
HR Lyr	18 h 53m 25.05s	+29° 13′ 37.8″	6.5–16.5	N/NR?	N/A		R
LL Lyr	18 h 35m 12.78s	+38° 20′ 04.7″	12.8–17.1	UG	05 h 59m		
V344 Lyr	18 h 44m 39.18s	+43° 22′ 28.2″	13.8–<20	UGSU	N/A		
V358 Lyr	18 h 59m 32.95s	+42° 24′ 12.2″	16.0–<20	N/UGWZ	N/A		R

(Continued)

Table 1.2. An alphabetical listing of 191 of the brightest or most easily observable cataclysmic variables, comprising mainly dwarf novae with a few novae and recurrent novae of interest thrown in (*Continued*)

Star	RA (2000)	Dec (2000)	V mag	Type	Orbit	Inc	R
V587 Lyr	19 h 17m 26.46s	+37° 10′ 40.8″	14.3–<17	UG	N/A		
CW Mon	06 h 36m 54.58s	+00° 02′ 17.2″	11.9–16.3	UGSS	04 h 39m	70°	
V616 Mon	06 h 22m 44.50s	–00° 20′ 44.0″	11.2–<20.2	HMXB	07 h 45m		R
RS Oph	17 h 50m 13.16s	–06° 42′ 28.5″	4.3–12.5	NR	456 days		R
V426 Oph	18 h 07m 51.69s	+05° 51′ 47.8″	11.5–19.4	UGZ/DQ	06 h 51m		
V2051 Oph	17 h 08m 19.09s	–25° 48′ 30.8″	13.0–17.5	UGSU	01 h 30m	83°	
V2204 Oph	18 h 26m 02.00s	+11° 55′ 06.0″	13.7–16.8	NL/ZAND	N/A		R
V2110 Oph	17 h 43m 33.38s	–22° 45′ 35.3″	12.0–20.0	NC	N/A		R
V2487 Oph	17 h 31m 59.8s	–19° 13′ 56.0″	9.5–17.7	N/NR?	N/A		
CN Ori	05 h 52m 07.79s	–05° 25′ 00.5″	11.9–16.3	UGZ	03 h 55m		
CZ Ori	06 h 16m 43.21s	+15° 24′ 11.4″	11.2–17.0	UG	05 h 15m		
V650 Ori	05 h 31m 08.82s	+09° 45′ 27.7″	15.5–<18	UGWZ	N/A		R
V1159 Ori	05 h 28m 59.55s	–03° 33′ 52.8″	11.2–15.1	UGSU	01 h 30m		
BD Pav	18 h 43m 11.90s	–57° 30′ 44.9″	12.0–<16.5	UG	04 h 18m	71°	
HX Peg	23 h 40m 23.70s	+12° 37′ 41.7″	12.9–16.6	UGZ	04 h 49m		
IP Peg	23 h 23m 08.53s	+18° 24′ 59.2″	10.5–17.8	UG	03 h 48m	81°	
RU Peg	22 h 14m 02.56s	+12° 42′ 11.5″	9.0–13.1	UGSS	08 h 59m		
FO Per	04 h 08m 34.99s	+51° 14′ 48.3″	11.8–16.2	UG	~04 h		
GK Per	03 h 31m 12.01s	+43° 54′ 15.4″	0.2–13.9	N/IP	47 h 55m		
KT Per	01 h 37m 08.79s	+50° 57′ 20.4″	10.6–16.1	UGZ	03 h 54m		
PY Per	02 h 50m 00.00s	+37° 39′ 22.0″	13.8–19.8	UGZ	03 h 43m		
TZ Per	02 h 13m 50.91s	+58° 22′ 51.9″	12.3–15.6	UGZ	06 h 19m		
UV Per	02 h 10m 08.32s	+57° 11′ 21.2″	11.7–17.9	UGSU	01 h 33m		
UW Per	02 h 12m 29.54s	+57° 05′ 18.5″	12.0–22?	N/NR?	N/A		R
V336 Per	03 h 22m 53.84s	+41° 37′ 01.4″	14.3–19.7	UG	N/A		R
V518 Per	04 h 21m 43.00s	+32° 54′ 00.0″	12.0–<20.0	N/XN	N/A		R
EI Psc	23 h 29m 54.21s	+06° 28′ 11.6″	12.7–16.6	UGSU	01 h 04m		R
XY Psc	01 h 10m 11.27s	+03° 32′ 33.9″	13.0–<18.5	UG	N/A		R
T Pyx	09 h 04m 41.51s	–32° 22′ 47.6″	6.3–15.3	NRB	01 h 50m		R
VY Scl	23 h 29m 00.48s	–29° 46′ 45.9″	12.9–18.5	VY	05 h 34m		
U Sco	16 h 22m 30.81s	–17° 52′ 44.1″	8.8–19.5	NRB	29 h 32m		R
V745 Sco	17 h 55m 22.25s	–33° 14′ 59.5″	11.2–21.0	NRA	N/A		R
V893 Sco	16 h 15m 14.98s	–28° 37′ 32.1″	10.6–14.5	UG	01 h 49m	70°	
EU Sct	18 h 56m 13.17s	–04° 12′ 33.4″	8.4–18.0	NR	N/A		R
FS Sct	18 h 58m 16.93s	–05° 24′ 05.2″	10.1–19.3	NR	N/A		R
NY Ser	15 h 13m 02.30s	+23° 15′ 08.4″	14.8–17.9	UGSU	02 h 21m		
QW Ser	15 h 26m 13.96s	+08° 18′ 02.3″	12.8–<15	UGSU	01 h 47m		
SW Sex	10 h 15m 09.40s	–03° 08′ 33.1″	14.8–16.7	NL	03 h 14m	79°	
AW Sge	19 h 58m 37.10s	+16° 41′ 28.6″	13.8–<18	UGSU	N/A		R
WZ Sge	20 h 07m 36.41s	+17° 42′ 15.4″	7.0–15.5	UGWZ/DQ	01 h 22m	76°	R
V729 Sgr	19 h 16m 49.10s	–26° 14′ 33.7″	14.1–15.5	UG	04 h 10m	70°	

(Continued)

Table 1.2. An alphabetical listing of 191 of the brightest or most easily observable cataclysmic variables, comprising mainly dwarf novae with a few novae and recurrent novae of interest thrown in (*Continued*)

Star	RA (2000)	Dec (2000)	V mag	Type	Orbit	Inc	R
V1017 Sgr	18 h 32m 04.47s	−29° 23′ 12.3″	6.2–14.8	ZAND	137 h 08m		R
V1172 Sgr	17 h 50m 23.55s	−20° 40′ 30.3″	9.0–18.0	N	N/A		R
V3645 Sgr	18 h 35m 49.21s	−18° 41′ 45.1″	12.6–18.0	NR	N/A		R
V701 Tau	03 h 44m 01.9s	+21° 57′ 08.4″	14.1–<21	UGSU	N/A		R
UW Tri	02 h 45m 17.29s	+33° 31′ 26.3″	14.7–22.6	UGSU/WZ	N/A		R
BZ UMa	08 h 53m 44.17s	+57° 48′ 40.6″	10.2–15.9	UGSU	01 h 38m		
CH UMa	10 h 07m 00.69s	+67° 32′ 47.3″	10.7–15.3	UG	08 h 14m		
CI UMa	10 h 18m 13.12s	+71° 55′ 44.2″	13.8–18.8	UGSU	01 h 26m		
DV UMa	09 h 46m 36.59s	+44° 46′ 44.7″	14.0–20.6	UGSU	02 h 04m	84°	R
DW UMa	10 h 33m 52.86s	+58° 46′ 54.7″	14.9–18.0	UX/VY	03 h 17m	80°	
ER UMa	09 h 47m 11.93s	+51° 54′ 09.0″	12.8–15.8	UGSU	01 h 32m		
IY UMa	10 h 43m 56.73s	+58° 07′ 31.9″	13.0–18.4	UGSU	01 h 46m	87°	
KS UMa	10 h 20m 26.52s	+53° 04′ 33.1″	13.0–17.0	UGSU	01 h 38m		
SU UMa	08 h 12m 28.27s	+62° 36′ 22.4″	11.2–15.0	UGSU	01 h 50m	80°	
SW UMa	08 h 36m 42.74s	+53° 28′ 38.1″	9.0–17.0	UGSU/DQ	01 h 22m		
UX UMa	13 h 36m 40.96s	+51°54′ 49.5″	12.7–14.1	UX	04 h 43m	71°	
SS Umi	15 h 51m 22.33s	+71° 45′ 12.0″	12.6–17.6	UGSU	01 h 38m		
HV Vir	13 h 21m 03.16s	+01° 53′ 29.1″	11.5–19.0	UGWZ	01 h 22m		R
OU Vir	14 h 35m 00.14s	−00° 46′ 07.0″	14.5–18.5	UGSU	01 h 45m	79°	
TW Vir	11 h 45m 21.16s	−04° 26′ 05.8″	12.1–16.3	UG	04 h 23m		
FY Vul	19 h 41m 39.95s	+21° 45′ 59.0″	13.4–15.3	UGZ	N/A		
SW Vul	20 h 00m 05.17s	+22° 56′ 06.9″	14.5–18.5	UG	N/A		
TY Vul	20 h 41m 44.0s	+25° 35′ 11.0″	14.0–19.0	UG	N/A		R
VW Vul	20 h 57m 45.07s	+25° 30′ 25.7″	13.6–15.6	UGSU	04 h 03m		

Data has been acquired from a variety of sources, not least the online Downes & Shara catalog and SIMBAD. The columns give the object names (listed in abbreviated alphabetical constellation order, except for the final New Suspected Variables, Lanning, Sloan, and other miscellaneous new objects in the lower section), their 2000 R. A. and Dec. positions, the magnitude range, the object type, and the orbital period (where known). Eclipsing systems are marked in a darker gray and the axial inclination is given, where known (90° is the perfect eclipser). A letter R in the final column indicates that the BAA included the star as worth monitoring for recurrent outbursts at the time of writing (2008). The object type abbreviations are as follows: AM=AM Herculis (Polar/synchronous rotators); AM CVn=AM Canes Ven., i.e., close white dwarfs such as CR Boo (a Helium star); CV=cataclysmic variable (precise type unknown); DQ=DQ Herculis subtype; HMXB = high-mass X-ray binary; IBWD = interacting binary white dwarf; IP = intermediate polar; N = nova; NR = recurrent nova; NL = nova-like variable; SW Sex = SW Sextans subtype; UG = U Gem subtype; UGSS = U Gem variable (SS Cyg subtype); UGSU = U Gem variable (SU UMa subtype); UGWZ = U Gem variable (WZ Sge subtype); UGZ = U Gem variable (Z Cam subtype); UX = nova-like variable (UX UMa subtype); VY = nova-like variable (VY Scl subtype, i.e., systems that undergo low states); XN = X-ray nova; Z AND = symbiotic variable (Z And subtype) column are some of the most enigmatic objects that are worth patrolling in case of a recurrent outburst. I have deliberately, if unconventionally, listed these objects in alphabetic constellation order rather than the usual R.A. order.

Not surprisingly the CVs that do outburst on an irregular basis or on a timescale of years, or even decades, are the 'holy grail' objects for the CV watchers. Spotting a rare outbursting CV carries a lot of kudos in the variable star-observing community, and these days you do not necessarily have to endure freezing cold conditions to spot an event. Using CCD cameras and a Go To telescope of even 100-mm aperture can prove to be a very effective CV patrol system. It is refreshing to see, in these days of mass-produced Schmidt–Cassegrains, that a few amateurs still use homemade telescopes for observing CVs and other objects, such as Denis Buczynski's system, shown in Figure 1.4. Even when amateur astronomers are clouded out there are robotic telescopes at remote locations that they use to carry out nightly checks on CVs. In the U.K. amateur astronomer Jeremy Shears has been very successful (using a 100-mm aperture Takahashi refractor and a Celestron 11 Schmidt–Cassegrain) patrolling, and carrying out photometry on, outbursting CVs. Of course, most newcomers to CV work will not detect any outbursts, but

Figure 1.4. Denis Buczynski with his fork mounted 33-cm f/3.5 Newtonian, which was used to obtain the IP Peg light-curve in the next figure.

they will be keen to observe CVs in outburst. This is where being part of an e-mail alert system is so important, and subscribing to the CVNet alerts is crucial.

The table is extended below to include some interesting, mainly new/suspected CVs, awaiting a constellation designation. See caption for abbreviated catalogue designations.

Star	RA (2000)	Dec (2000)	V mag	Type	Orbit
NSV 00895	02h 42m 09.10s	+43° 21′ 08.0″	11.7–<20.0	UG	N/A
NSV 18241	09h 38m 36.98s	+07° 14′ 55.1″	12.9–16.8	UG	N/A
NSV 24587	18h 46m 51.10s	−04° 56′ 38.0″	8.0–22.0	UG?	N/A
NSV 25747	21h 42m 58.96s	+31° 34′ 42.7″	12.8–<17.0	UG	N/A
NSV 25966	22h 50m 39.64s	+63° 28′ 39.3″	16.5–N/A	NL	N/A
Lanning 17	18h 23m 01.20s	−04° 37′ 19.0″	15.0–20.0	NL	N/A
SDS0729+36	07h 29m 10.68s	+36° 58′ 38.3″	N/A–20.6	NL	N/A
SDS0747+42	07h 47m 16.81s	+42° 48′ 49.0″	N/A–16.8	NL	N/A
SDS0804+51	08h 04m 34.20s	+51° 03′ 49.2″	12.0–18.0	UGWZ	N/A
SDS0901+48	09h 01m 03.94s	+48° 09′ 11.1″	16.1–19.3	CV	~1.9 h
SDS2303+01	23h 03m 51.64s	+01° 06′ 51.0″	N/A–17.5	UG	N/A
FBS1719+834	17h 13m 06.34s	+83° 18′ 14.0″	14.0–<20.0	UG	N/A
FBS1735+825	17h 29m 42.80s	+82° 26′ 55.2″	14.0–<20.0	UG	N/A
FSVJ1722+2723	17h 22m 43.96s	+27° 23′ 55.7″	N/A–<21.0	UGWZ	N/A
EUVE J0854+39	08h 54m 13.90s	+39° 05′ 37.2″	N/A–16.4	AM Her?	N/A
1RX0532+62	05h 32m 33.87s	+62° 47′ 52.1″	11.9–16.4	CV	N/A

Extended table for newer objects awaiting catalog designations, with abbreviated identifiers: NSV = new suspected variable; Lanning = Palomar Schmidt objects spotted by Howard H. Lanning; SDS (abbreviated to fit column) stands for SDSSJ (Sloane Digital Sky Survey, J Filter); FBS = First Byurakan Survey; FSV = Faint Sky Variability Survey; EUVE = extreme ultraviolet explorer; 1RX (abbreviated to fit column) stands for 1RXS = Rosat X-ray Source Catalogue

Worthwhile Projects

Detecting Eclipsing Dwarf Novae

With more than 600 dwarf novae (classified as UG or UG subtypes) known, and more than 900 CVs in the definitive Downes & Shara catalog (excluding novae and nova-like variables), one might expect a huge number of them to exhibit eclipses (refer again to the lower diagram in Figure 1.1). After all, dwarf novae are compact systems with both stars and the white dwarf accretion disk typically occupying a volume of space similar to that of our own Sun. According to Bill Worraker, of the British Astronomical Association, about 34 percent of dwarf novae should show evidence of some sort of eclipse activity in their quiescent light-curves, that is, the faint red dwarf eclipsing the accretion disk (and hot spot) when the orbital inclination is 70° or higher.

For even higher orbital inclinations the white dwarf itself will be eclipsed. Obviously those systems that outburst regularly provide far more light to work with when looking for eclipses, but most of the time dwarf novae are not in outburst. Those dwarf novae that outburst regularly and have favourable orbital inclinations have provided astronomers with most of their data on how these systems work (see Figure 1.5 for an eclipse light-curve of IP Peg by Denis Buczynski). So, with more

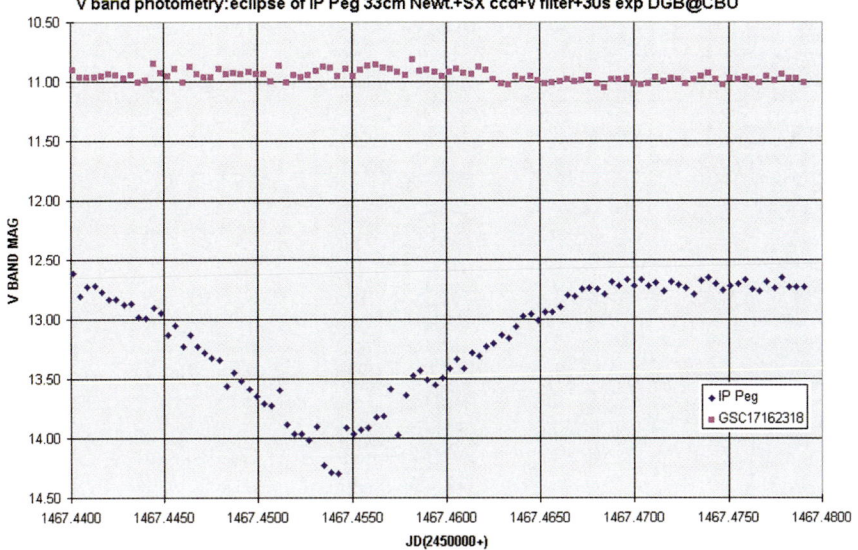

Figure 1.5. A light-curve of IP Peg obtained by Denis Buczynski on October 15, 1999, with his 33-cm f/3.5 Newtonian showing an eclipse of the hot spot by the dim red secondary star.

than 600 dwarf novae known, and 34 percent theoretically capable of revealing some eclipse activity (15 percent should reveal substantial eclipses), how many eclipsing dwarf novae are actually known? Do not get too excited! The current number is around 20, depending on precisely how you define a dwarf nova in the borderline cases. Obviously something is wrong here, and the most likely reason for the factor-of-10 discrepancy in the numbers of eclipsing dwarf novae detected (compared to that predicted) is surely a lack of monitoring by skilled photometrists. However, in recent years the BAA's Variable Star Section started a campaign (led by Bill Worraker) to discover more eclipsing dwarf novae, and so far, no more have been discovered. The rows in Table 1.2 show a darker font for systems where eclipses take place; the orbital inclination (where known) in such cases is indicated, too. The study of these valuable eclipsing systems, especially when the dwarf nova is in outburst, has produced vital photometric data that helps in the study of all these objects. For dwarf novae where a semipredictable outburst period is known (i.e., the accretion disk flares on a regular, mostly reliable basis), the star is especially important because amateurs and professionals can be fully prepared for combined outbursts and eclipses, and the more outbursts covered, the more the system is understood. The systems U Geminorum and IP Peg are especially valuable in this regard, but a few other systems are known eclipsers and outburst on a semiregular basis, providing a veritable gold mine of data for the theorists.

Table 1.3 details the most valuable systems in this regard, that is, the systems with the shortest predictable periods between bright outbursts. U Gem and IP Peg show especially bright outbursts, but Z Cha and OY Car (both in the far south), IY UMa, LBQS 1432-0033, and HT Cas are very convenient to monitor due to their short orbital periods. When a dwarf nova is below the period gap a full light-curve can be captured in less than 2 hours.

But we have digressed here. Only 20 or so eclipsing dwarf novae are known, and despite efforts to find more, by patient amateur monitoring of the established systems,

Table 1.3. The most valuable eclipsing dwarf novae with semipredictable outbursts, in rough order of importance to the theorists (i.e., the most predictable bright and fast-orbiting dwarf novae rank highest)

Dwarf Nova	Mag. range	Outburst interval	Orbital period	Inc.
U Gem	9.1–15.2	100 days	04 h 15m	70°
IP Peg	10.5–17.8	95 days	03 h 48m	81°
Z Cha	11.9–17.2	51 or 218 days	01 h 47m	82°
IY UMa	13.0–18.4	90 or 255 days	01 h 46m	87°
LBQS 1432-0033	14.5–18.5	80 or 410? days	01 h 45m	N/A
OY Car	12.4–17.3	160 or 346 days	01 h 31m	83°
HT Cas	10.8–18.4	400 days?	01 h 46m	81°
EX Dra	12.5–17.2	Approx 20 days	05 h 02m	86°
EM Cyg	12.5–14.5	Approx 20 days	06 h 59m	67°
DV UMa	14.0–20.6	Approx 700 days	02 h 04m	84°
BD Pav	12.0–<16.5	>1000 days?	04 h 18m	71°
WZ Sge	7.0–15.5	11,786 days	01 h 22m	76°

Additional data and many more eclipsing dwarf novae can be found in Table 1.2

no more have been found by backyard photometrists in recent years (but read on to the end of this section). Specifically, during Bill Worraker's own BAA project the dwarf novae AR And, FO And, KV And, TT Boo, AT Cnc, CC Cnc, GX Cas, KU Cas, V516 Cyg, ES Dra, AW Gem, SX LMi, CY Lyr, V344 Lyr, V426 Oph, TZ Per, CI UMa, HS Vir, and VW Vul have had partial-orbit or full-orbit photometry carried out, and there has been no evidence of any eclipses. The current (2008) program includes some of these stars and others that amateurs are encouraged to monitor over an entire orbit. Specifically, the new list includes AR And, FO And, KU Cas, SV CMi, V516 Cyg, V1060 Cyg, ES Dra, CY Lyr, LL Lyr, V426 Oph, HX Peg, PY Per, and FY Vul (Table 1.4).

Amateurs with the ability to do accurate photometry down to magnitude 17 or so (unfiltered photometry is fine for this work) are encouraged to monitor these

Table 1.4. Some CV systems in Bill Worraker's BAA program that are thought to be worth monitoring for eclipse activity

STAR	RA (2000)	Dec (2000)	Range	Type	Orbit
AR And	01 h 45m 03.28s	+37° 56′ 32.7″	11.0–17.6	UG	03 h 55m
FO And	01 h 15m 32.14s	+37° 37′ 35.6″	13.5–17.5	UGSU	01 h 43m
KU Cas	01 h 31m 02.37s	+57° 54′ 13.4″	13.3–19.7	UGSS	N/A
SV CMi	07 h 31m 08.40s	+05° 58′ 48.3″	13.0–16.9	UGZ	03 h 45m
V516 Cyg	20 h 47m 09.78	+41° 55′ 26.5″	13.8–16.8	UGSS	N/A
V1060 Cyg	21 h 07m 42.18s	+37° 14′ 08.8″	13.5–18.0	UGSS	N/A
ES Dra	15 h 25m 31.79s	+62° 01′ 00.2″	13.9–16.3	UGSU?	~3 h?
CY Lyr	18 h 52m 41.38s	+26° 45′ 31.5″	13.2–17.0	UG	03 h 49m
LL Lyr	18 h 35m 12.78s	+38° 20′ 04.7″	12.8–17.1	UG	05 h 59m
V426 Oph	18 h 07m 51.69s	+05° 51′ 47.8″	11.5–19.4	UGZ/DQ	06 h 51m
HX Peg	23 h 40m 23.70s	+12° 37′ 41.7″	12.9–16.6	UGZ	04 h 49m
PY Per	02 h 50m 00.00s	+37° 39′ 22.0″	13.8–19.8	UGZ	03 h 43m
FY Vul	19 h 41m 39.95s	+21° 45′ 59.0″	13.4–15.3	UGZ	N/A

stars. In a nutshell, 19 dwarf novae have been monitored so far, 11 over one or more orbital periods, and yet there has been no sign of eclipses. So what has happened to this 34 percent statistic? How has it become 0 percent for these 11 objects? A good question!

At the time of writing it is a bit of a mystery, but one that the well-equipped amateur can have a go at solving. It should be stressed that although dwarf novae in outburst are far easier to check out for eclipses (because they are so much brighter), using CCDs and modest-aperture telescopes eclipses can be detected by amateurs without the system being in outburst. Of course, much depends on the specific star and the aperture of the telescope, but if you can image the star easily, at its minimum brightness, you can detect eclipses. If you are interested in testing your photometric system on known eclipsing CVs there are various websites that contain predictions. At the time of writing, British amateur Gary Poyner's website contained a page of links to such predictions and can be found at http://www.garypoyner.pwp.blueyonder.co.uk/cveclipse.html.

The many CVs that are being discovered by the Sloane Digital Sky Survey (SDSS) (28 in 2005 alone) may also bear fruit in this regard. Indeed, in October 2007 U.S. amateur Steve Brady announced that he had imaged the Sloan Deep Sky Survey object cataloged as SDSS J090103+480911 in outburst at magnitude 16.1 and that several hours of time-series observations revealed 'eclipses with a period of $\sim$1.9 hours and an eclipse depth of $\sim$1.0 magnitudes'. So, there are new eclipsing systems out there even if, statistically, there should be a lot more.

Dwarf Nova Monitoring

The outbursts of dwarf novae are rarely predictable with any great certainty. Yes, there are major exceptions to this sweeping statement, but, in the main, and especially with the rarest outbursters, they need to be monitored every clear night, or an outburst may be missed. Part of the problem is that when outbursts occur separated by years there simply is not enough data to accurately predict future activity. Even the most dedicated observers living in the sunniest climates will not check their favorite objects every night for a year. For the visual observer the presence of a nearby moon or a near full Moon will write off many days each month. Pushing a telescope to point at numerous faint targets every night is hard work, but since the mid-1990s the combination of affordable Go-To telescopes and CCD cameras has meant that checking galaxies for supernovae and checking dwarf nova fields for outbursts has become less onerous, even if it is tougher on the wallet and there is a learning curve. CCDs have also enabled fainter dwarf nova outbursts to be detected, even with relatively small telescopes. There is useful work to be done here, and backyard observers are monitoring both newly discovered dwarf novae and rare outbursting old favorites every clear night. Indeed, one of the major organizations involved in the photometry of CV stars is actually called the CBA (Center for Backyard Astrophysics). They have a website at http://cba.phys.columbia.edu/. Their single aim is producing light-curves for variable stars using photometry.

The CBA was originally founded in the 1970s by David Skillman, a U.S. amateur who was about 20 years ahead of his time. He described his automated photometry system in the popular U.S. astronomy magazine *Sky & Telescope* in January 1981. His 32-cm reflector attracted much attention, as it could be left unattended to collect light-curves of variable stars during the night, even while he was asleep. In the 1980s a similar system was built by the Crayford (British) amateur astronomer Jack Ells. His local astronomical society was the Crayford Manor House Astronomical Society (CMHAS), which may be a familiar name to many amateurs; as a fellow CMHAS member, John Wall invented the well-known 'Crayford Focuser' now used worldwide.

Of course, in 2008, automated systems are commercially available, as is the necessary slewing and photometry software. However they were unique and revolutionary prior to the 1990s. Skillman's telescope used an IP21 photomultiplier tube to collect the photons, but in the 1990s he and other amateurs started switching to CCDs that provided an extra 3 magnitudes of sensitivity. Suddenly photometry could easily monitor stars fainter than the human eye could see. The CBA (initially called the Center for Basement Astrophysics) sprang from Skillman's pioneering robotic photometry, and it soon acquired new members, with similar equipment, across the world. This was extremely useful, especially when there was a need to monitor CVs with long orbital periods over many hours or, simply, to minimize the effects of cloud cover. In 2008 the CBA has roughly 50 active members who can acquire good photometric data using 20-cm and larger telescopes. Their members are mainly spread across the U.S., Western Europe, Australia, and New Zealand, but a few other CBA outposts also exist such as at supernova discoverer Berto Monard's observatory in South Africa.

The CBA and similar organizations, like the AAVSO and BAA VSS (British Astronomical Association Variable Star Section) tend to concentrate on the two most vital areas of dwarf nova research, namely, actually patrolling for outbursts, and the monitoring of 'superhumps' in the light-curve. In addition, as discussed in the previous section, discovering more eclipsing dwarf novae, by photometry of less well studied CVs, is a more recent sideline.

Back to our main subject. In 2008 all the major telescope manufacturers were offering 'Go To' drives on their higher spec models that enable keypad slewing to precise positions. In general, the more expensive the Go To systems, the more likely they are to hit the target. The cheapest systems, using wormwheels smaller than 100 mm in diameter, will often miss the target by 10 arcminutes or more, despite advertising hype telling you otherwise. The instrument's field of view should match the slewing accuracy of the Go To system employed.

However, even when precision slewing is available there is another consideration. Although Messier, NGC, SAO, and even Patrick Moore's Caldwell designations are often included in popular telescope databases, hardly any dwarf novae will appear, unless, like U Gem, they fall under the category of a bright variable star. What is needed is the ability to create a customized script file where dozens of rare outbursting dwarf novae can be named and recognized by the software that controls the telescope and where, immediately after each slew, a CCD image is taken of the field. Fortunately, the software suite of The Sky, CCDSoft, and Orchestrate, by Software Bisque enables this to be done. We will have more to say about the integration of these packages for supernova hunting in Chapter 4.

Using Robotic Telescopes for Dwarf Nova Monitoring

It might be thought that to systematically patrol dozens of CVs each night from an armchair, you would need to invest $20,000 or more in a quality 'Go To' telescope, a CCD camera, and some good software. Well, things are not quite as bad as that. (Indeed, if you dispense with the armchair and observe visually, a high-quality 0.4-m Dobsonian and some digital setting circles can be acquired for well under $3000). If you go down the CCD imaging route even a 100-mm-aperture telescope can be pushed to magnitude 18 (far deeper than any visual observer can reach, even with a 0.45-m Dobsonian) and with quality, low-cost, Chinese Go To mounts, such as the Synta EQ6 Pro, using Go To need not break the bank.

However, there is another route that is becoming increasingly popular among students, schools, and simply observers who do not have access to their own observatory. Indeed, even keen observers who do have their own telescopes are using this route when they are clouded out, or an object is below their horizon/tree line. The solution is using an online robotic telescope via a web browser. In recent years there has been a surge in the use of robotic telescopes as educational tools, for schools and colleges, or for use by university departments wishing to access darker and clearer skies at more favorable latitudes. In the U.K. there are three major Robotic Telescope Projects, and these are the Bradford Robotic Telescope (BRT), run by Bradford University, the Faulkes Telescopes Project, and the Liverpool John Moore's University National Schools Observatory project. There are also similar educational projects in the U.S. The BRT consists of that proven optical package, a Celestron C14 (355-mm f/11 Schmidt–Cassegrain), mounted on that proven robotic mount, the Paramount ME. The telescope (www.telescope. org) is on the Observatorio del Teide site of the Instituto De Astrofisica De Canarias in Tenerife, the Canary Islands, Spain. The Teide Observatory is the best in Europe and is situated at an altitude of 2400 m (7900 feet) on the northern part of the volcano caldera. At the time of this writing, 2008, the telescope had completed more than 14,000 image requests, and in excess of 20,000 users were registered with the telescope site. The two Faulkes telescopes (http://faulkes-telescope.com/) are a considerably bigger technological investment, as they are substantial 2-m-aperture instruments based in Hawaii and Siding Spring Observatory, Australia. The Faulkes Telescope Project was launched in March 2004, with funding from the Dill Faulkes Educational Trust, established by business entrepreneur Dr. Martin 'Dill' Faulkes. The Faulkes telescopes were designed and built by Telescope Technologies Limited, a spin-off company from Liverpool John Moores University in northwest England. (TTL is now a wholly-owned subsidiary of the Las Cumbres Observatory Global Telescope Network, or LCOGTN). The 2-m Liverpool Telescope (http://telescope.livjm.ac.uk/), based at the Observatorio del Roque de Los Muchachos on the Canary island of La Palma, Spain, was TTL's first 2-m instrument and became actively robotic in 2004, paving the way for TTL's Faulke's instruments, an online educational facility for schools, and a future 'ROBONET' project to build up to seven 2-m telescopes around the world for nonprofit educational research. Of course, it goes without saying that access to observing time on 2-m telescopes is not that easy for amateurs to acquire, but if a sensible proposal is put forward, anything can happen. The German amateur Patrick Schmeer has successfully used the online robotic University of Iowa 'Rigel Telescope' to catch a number of CV outbursts. The main advantage of all these university-based robotic telescopes is that they operate on a zero-profit basis with schools and make educational use a top priority.

However, money talks, and there are also various facilities springing up around the globe offering telescope time for your dollars, pounds, or yen. One of the best-equipped facilities for online amateurs with a budget to spend on robotic observing is the New Mexico Skies iBisque collaboration (http://www.nmskies.com/webpage/ibis-que.html), where amateurs can take advantage of the dark and clear New Mexican night skies as well as Software Bisque's expertise in online robotic control; subscribers can rent time on 12 Celestron 14s mounted on Bisque's renowned Paramount ME mounting. Telescope time can (at the time of this writing) typically be paid for at between $17 and $50 per hour. The 'Global Rent-a-Scope' organization (www.global-rent-a-scope.com) also uses the New Mexico Skies facilities and has five telescopes (Takahashi Mewlon Dall-Kirkhams and a Takahashi Apochromat) in New Mexico as well as a telescope in Israel and two more in Australia. One way to access time on these instruments is via RASO (the Remote Astronomical Society Observatory) at www.remote-astronomical-society.org/.

Another system, based on Mt Teide in Tenerife, can be found at www.slooh.com. At the time of writing this 'Slooh' internet observatory was planning two more online facilities in Chile and Australia, ensuring 24-hour access to the night sky.

Online robotic telescope usage is nowhere near as convenient as having your own telescope for CV monitoring in your backyard, and you can never put your eye to the eyepiece, but it is a solution if you live in a city, an apartment, or have endured several weeks of solid cloud, which is quite typical in some places.

CV Outburst Successes

For many years the BAA VSS, in close cooperation with Guy Hurst's *The Astronomer* magazine, has operated a Recurrent Objects Program (ROP) that consists of a table of poorly observed dwarf novae along with suspected recurrent novae and nova-like variables. Put another way, it contains objects of type UG, UGSU, UGSU+E, UGWZ, N, NR, NL, VY, IP, or ones suspected of belonging to these categories. This table is under continuous revision, as once an object's outburst period is proven, which may take years or even decades, it can be dropped from the list. Conversely, as new objects are discovered, either by amateur nova patrollers or professional surveys, they can be added to the list. If you refer back to Table 1.2 and look at the final column, an 'R' in that column indicates that the object is currently (2008) on the BAA VSS 'ROP' list. Obviously most of these objects are well within the Milky Way, and there are quite a few toward the galactic center located in Sagittarius. Further from the visible river of the Milky Way these objects are still within our galaxy, of course, but they tend to be closer to us and simply above or below us with respect to our galaxy's 2000-light-year-thick outer disk.

It goes without saying that there is a lot of pride and kudos associated with being the first person to detect a really rare outbursting CV going off, and there have been some very enigmatic objects on the ROP list. Admittedly the greatest prestige is usually associated with detecting the outbursts of recurrent novae (see Chapter 2), like RS Ophiuchi, but there are quite a few dwarf novae that only outburst a few times in an observer's lifetime. This is especially true when an observer lives in a cloudy climate such as the U.K., where the most overcast periods usually coincide with the most spectacular astronomical events.

Checking the objects in the BAA VSS ROP list has become a truly international project in recent years (and other organizations now have similar lists). Although the project was started by Guy Hurst in the mid-1980s, it was largely inspired by the remarkable amateur (as he was then) astronomer Rob McNaught, who has become the world's greatest comet discoverer in recent years as well as being a nova and

Figure 1.6. The UGWZ-type dwarf nova VY Aquarii in outburst on October 7, 2006, in full moon-light! A 120-second exposure by the author using a 0.35-m Celestron 14 Schmidt–Cassegrain working at f/7.7, atop a Paramount ME mount. An SBIG ST9XE CCD was used for this 13 arcminute square image with north at the top.

supernova discoverer! McNaught searched historic Sonneborg Observatory images of the field around the recurrent nova (as it was then classified) VY Aquarii that had flared up in 1907 and 1962 (see Figure 1.6). He discovered other outbursts and then, on November 28, 1983, he made the first-ever visual observation of VY Aqr in outburst. Several further outbursts have since been witnessed, and, as a result, VY Aqr has been reclassified as a UGWZ dwarf nova with an orbital period of 1 hour 31 minutes and an outburst range from 17.5 to 8.0. Shortly afterward (November 1, 1985) U.K. amateur Stephen Lubbock discovered another object on McNaught's list in outburst, namely DO Draconis, and following these successes Guy Hurst drew up a list of some 75 stars worth monitoring for further outbursts. Stephen Lubbock also discovered RZ Leonis in outburst on November 28, 1987.

The years 1991–1993 were especially productive for the ROP program, operated jointly between the BAA VSS and Guy's magazine *The Astronomer*. During that time the first-ever visual observations of SS UMi, EF Peg, HV Vir, AK Cnc, DI Uma, V1113 Cyg, S10930 Lyr, and LL And were bagged by patrol members.

The 12th-magnitude outburst of HV Vir, discovered by German amateur Patrick Schmeer, was especially memorable. HV Vir was photographed by Schneller from Babelsberg Observatory in 1929 (February 1st and 3rd) but had faded and was too faint to be followed by February 9th. No spectroscopic confirmation of its nature was obtained, but the fact that no star brighter than magnitude 19 was known at that location implied the object was a nova. However, Schmeer's discovery of the star in outburst again, at 12th magnitude, on April 20, 1992, enabled professional astronomers to get photometric and spectroscopic results showing that HV Vir was in fact a dwarf nova, exhibiting many characteristics of the rare WZ Sge objects as well as the superhumps more usually associated with UGSU stars.

QY Per, DV Uma, and UW Tri were also bagged in outburst by Tonny Vanmunster, observing from Belgium, on October 25, 1994, February 26, 1995, and March 3, 1995, respectively. Vanmunster patrolled dwarf novae candidates on the Belgian VVS CVAP program, which is a similar project to the BAA VSS ROP program.

Another memorable and rare CV outburst was that of AL Com on April 5, 1995. This author was lucky to have a clear sky on the following night that enabled him to image the event (see Figure 1.7). The outburst was spotted by David York of the U.S., who observed it at magnitude 14.9. Eleven hours later Tom Cragg estimated the magnitude as 12.7. The rest magnitude of AL Com is magnitude 20.8, so it had undergone a nova-like increase in magnitude. AL Com lies in a very distinctive field; it is only 8 arcminutes southeast from the well-known galaxy M88 (NGC 4501). So it is not inconceivable that supernova patrollers with a wide enough field could detect further outbursts. Prior to April 5, 1995, the previous known outbursts were on November 17, 1961, and in April 1974 and March 1975. A fifth outburst occurred on May 18, 2001, and a sixth (poorly placed in the morning sky) in October 2006. Amateur and professional observations of AL Com have since shown that it is a rare WZ Sge-type dwarf nova, that is, a subclass of the SU Uma group of CVs. The system's orbital period is 82 minutes, according to the Catalog and Atlas of Cataclysmic variables by Downes & Shara.

Nineteen months after the AL Com outburst, on November 30, 1996, Patrick Schmeer visually detected a 12th-magnitude outburst of the dwarf nova EG Cancri that had not been seen in outburst since November 1977. The detection had amateur and professional observers scrambling to obtain good time-series photometry, that is, photometry over periods of hours to extract vital orbital information.

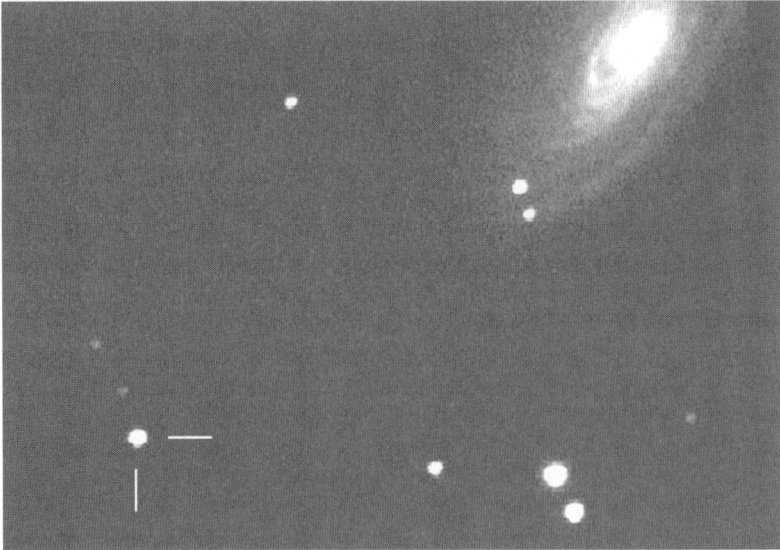

Figure 1.7. A very rare outburst of the UGWZ-type dwarf nova AL Coma Berenices, which lies between us and the much more distant galaxy M88. Image by the author on April 6, 1995, using a 0.49-m f/4.5 Newtonian and Starlight Xpress CCD.

On December 20, 2000, Rod Stubbings of Drouin, Victoria, Australia, detected a rare outburst of RZ Leonis, whose last recorded outburst had been on November 28, 1987. RZ Leonis was originally discovered by the legendary Max Wolf at Heidelberg on March 13, 1918, and was presumed to be a nova, but it is now known to be another rare dwarf nova of the SU UMa WZ Sge subclass.

On July 23, 2001, the Japanese observer Oshima was the first to spot the long-awaited outburst of WZ Sge itself, the defining member of its subtype. The star rose to magnitude 9.5. Previous outbursts had first been observed on November 22, 1913, June 28, 1946, and December 1, 1978.

The monitoring of alleged 'rare' outbursters by the BAA VSS ROP has shown that, in some cases, the only reason that outbursts were assumed to be rare was because they were simply not monitored regularly. SS UMi and V344 Lyrae were in this category but have now been proved to outburst on a regular basis and to be very interesting short-period UGSU stars. Conversely HN Cyg, which was originally classed as a U Gem dwarf nova in the GCVS (General Catalogue of Variable Stars), was shown to be a semi-regular (SR) star due to ROP surveys.

In 2008 many objects on the ROP list are being checked visually by keen observers every clear night. The world's greatest CV observer, Gary Poyner (we shall meet him later), can see to magnitude 16.8 with a 35-cm Schmidt–Cassegrain! Nevertheless, CCDs can go much fainter and carry out precise photometry on stars that can only be glimpsed visually. So, with many observers being equipped for digital imaging, it is possible to monitor much fainter stars for outbursts that are right on the limit, or below the limit, for most amateurs, that is, outbursts of magnitude 15 and fainter. The BAA has a CCD Dwarf Nova list specifically for such objects, and some of these are described below. We have provided some thumbnail images of the 5-arcminute fields for these stars (except V455 And and V478 Her, which have figures in this chapter) in the Resources section of this book.

Some Good Dwarf Nova Targets for CCDs

V402 And
00 h 11m 07.30s +30° 32′ 36.0″

On September 20, 2000, this author detected a rare outburst of the dwarf nova V402 And when he imaged the field with a 0.49-m Newtonian and a 120-second CCD exposure. He immediately phoned the U.K. coordinator Gary Poyner, reporting a (unfiltered) magnitude estimate of 15.7. The skies were clear in Birmingham, and, remarkably, Gary was able to confirm the outburst visually with a 0.4-m Dobsonian. He estimated the star as slightly brighter, at magnitude 15.4. Only 2 days earlier the Belgian observer Tonny Vanmunster had imaged the field to magnitude 17.6, and the star was not recorded. Gary alerted observers worldwide to my detection and, less than a day later, Japanese observers at Ouda observatory had carried out photometry, detected superhumps in the light-curve, and measured a superhump orbital period close to 1 hour 30 minutes. All in all the author was pretty satisfied with my outburst detection! This faint and rare outbursting dwarf nova is well worth keeping an eye on, especially if you have CCD equipment.

V455 And (formerly HS2331+3905)
23 h 34 m 01.55s +39° 21′ 42.9″

As this author was working on the first draft of this very chapter, in early September 2007, one of the most memorable CV outbursts of recent years took place in the constellation of Andromeda, about 1 hour west, in R.A. star-hopping terms, from the position of the well-known Andromeda galaxy, M31. In recent years the Hamburg Quasar survey has picked up a number of emission-line objects that are not at cosmological distances of billions of light-years, but are actually dwarf novae within our own galaxy. These objects are invariably given a provisional R.A. and Dec. designation prefixed by 'HS.' One of the most exciting Hamburg Survey finds in recent years was the CV designated HS2331+3905, which, even before it went into outburst, was being called 'The cataclysmic variable that has it all.' In the U.K. Boris Gaensicke of Warwick University and the prolific visual CV observer Gary Poyner were giving it top priority in May and June 2007, with local TV coverage even mentioning the pro-am collaboration on this star.

Why did HS2331+3905 look so exciting? Well, for starters, it has a very short orbital period, not much above the 78-minute limit we mentioned before. Early measurements indicated an 81-minute orbital period. In addition the white dwarf itself appeared to be pulsating on a timescale of 5–6 minutes and rotating rapidly, namely, once every 67 seconds! Professional astronomers thought the white dwarf was at a temperature of around 10,500 K, close to the ZZ Ceti instability region, at approx. 11,000 K, where white dwarfs can pulsate due to partially ionized hydrogen. In addition it appeared that the secondary might be a brown dwarf. It also seemed to have a favorable orbital inclination, enabling grazing eclipses to take place and so much science deduced. The star also appeared to have permanent superhumps, even when not in quiescence, and its spectra showed a 3.5-hour period radial velocity (from Balmer emission lines) that seemed to have no connection to its orbital period.

But perhaps most exciting of all was the apparent distance of the system. It appeared to be only 290 light-years away, making it potentially a binocular object if it were to experience a full outburst. Such an event, to 8th mag or so, would put HS2331+3905 into a very exclusive dwarf nova club. Only U Gem, SS Cygni, VY Aqr, GW Lib, RU Peg, WZ Sge, and SW Uma were proven to be in that brightness league, unless recurrent novae are included. The British observer Gary Poyner was keeping track of the object every clear night in Summer/Autumn 2007 in case it went into outburst. When it was cloudy he was using the online BRT (at Mt Teide, Tenerife) to monitor the star. On September 3, 1992 UT Gary made a CCD observation with the BRT showing HS2331 at minimum at 16.24 C. On the morning of September 4th it was clear at the BRT site, but when Gary logged on the software said the observatory could not be opened due to high wind speeds. This was a case of Murphy's Law in action, as that was the day the star decided to go into outburst. The Japanese observer Hiroyuki Mehara spotted it first, at magnitude 14.6. Every keen CV watcher on Earth then turned their attention to the suspected UGWZ star. The magnitude peaked at around 8.2 on the night of September 5th/6th, and huge amounts of valuable data were secured by amateurs and professionals alike. Figure 1.8 shows an image by the author of the star on September 4th, rising to a maximum the next night. A light-curve of the superhumps recorded by Gianluca Masi appears in Figure 1.9. Following the outburst this extraordinary object was promoted; it was given the sensible, non-tongue twisting constellation designation of V455 Andromedae.

Figure 1.8. The recently discovered "cataclysmic variable that has it all," HS2331+3905, designated V455 Andromedae in 2007, went into outburst in September 2007 and was caught here, on the rise, by the author. In this image, taken on Sept 4.917 UT, it was 12th mag. It reached 8th mag the next night, rivaling SS Cyg and U Gem's claims for the brightest/easiest to observe dwarf nova! (Celestron 14 at f/7.7 plus SBIG ST9XE CCD. 120 seconds. North at top).

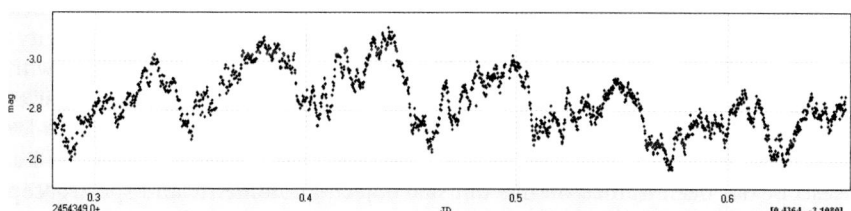

Figure 1.9. Gianluca Masi of Bellatrix Observatory captured this dramatic light-curve of superhumps from HS2331+3905/V455 Andromedae on the night of September 5/6, 2007. The horizontal axis spans 8 hours and the vertical axis 0.7 magnitudes (i.e., a light output doubling). 36-cm f/8 Schmidt–Cassegrain + SBIG ST7 CCD. Image by kind permission of Gianluca Masi/Bellatrix Observatory.

1502+09 Boo = EQ 1502+09
15 h 04 m 41.78s +08° 47′ 54.1″

Another faint target for CCD observers, this is possibly the least-observed object in this brief descriptive list. Discovered in the early 1980s on a 1-hour photographic exposure with the 1.2-m U.K. Schmidt (plus objective prism), this magnitude-18.6

recently been dropped from the BAA ROP program due to the successful observation of seven outbursts in 2005. Analysis of these observations showed CG Dra was rarely at quiescence and had an outburst period of around 11 days. Two types of outburst were identified: short ones lasting about 4 days and long ones lasting about 8 days. Nevertheless, the star is highly active and still worthy of further study.

DV Dra
18 h 17 m 23.10s +50° 48′ 18.1″

This dwarf nova, classified as a UGWZ type, is a prime target for CCD monitoring, as it only peaks at magnitude 15 and has a low state of around magnitude 21. CCD monitoring has revealed it experiences a number of faint outbursts to about magnitude 17, which are way too faint for even the keenest visual patrollers, even with 0.4-m or 0.45-m instruments. Superhumps have been seen, too. DV Dra is a challenging object but one that few others will be patrolling.

ES Dra
15 h 25 m 31.79s +62° 01′ 00.2″

A third high-declination dwarf nova in Draco, this CV will be easily circumpolar from many north European and North American amateur observatories the whole year round, so any outburst is unlikely to be missed. The star is still on Bill Worraker's BAA VSS eclipsing dwarf nova lists, too, as despite there being no evidence of eclipses in recent photometry some 'flickering' activity was recorded. Superhump outbursts have been suspected, but these need confirming. Pinning down the exact nature of this dwarf nova has proved elusive, too. Until recently the orbital period was thought to be 4 hours 18 minutes, but this has now been revised downward to roughly 3 hours, that is, just above the 'period gap.' The original UGSU classification of this dwarf nova is certainly in considerable doubt, so careful monitoring of it, and more observations in outburst, are required. With a magnitude range of 16.3–13.9 this is not exclusively a CCD target.

KV Dra
14 h 50 m 38.31s +64° 03′ 28.4″

Yet another very high-declination dwarf nova in Draco, this object shows significant low-level activity in the 16th-magnitude range as well as frequent short outbursts. It has been seen as bright as magnitude 11.8, but more observations are required.

CI Gem
06 h 30 m 05.86s +22° 18′ 50.7″

The northern hemisphere winter constellation of Gemini contains the famous U Gem, the first dwarf nova to be discovered, as well as a few fainter examples like this one. Cl Gem is very faint at minimum and only reaches magnitude 14.7, at best, in outburst. Its type classification is still in some doubt as is its orbital period. This is a good target for CCD monitoring, and time-resolved photometry when in outburst. The April 2005 outburst produced some photometry, but, low down and at 16th magnitude, its precise nature is still uncertain.

V478 Her
17 h 21m 05.60s +23° 39′ 36.8″

This author imaged V478 Her on April 27, 2006 (see Figure 1.10), following the detection of its outburst by U.K. dwarf nova observer Jeremy Shears. The star normally sits in quiescence at around magntude 17, but Jeremy captured it at 16.1. The object had only been in outburst on one previous occasion, in June 2001, when the Belgian observer Tonny Vanmunster detected superhumps indicating that V478 Her was a UGSU star. On that occasion the star had a superoutburst and peaked at magnitude 15.5. V478 Her was actually brighter than normal for several months in mid-2006, but it is not known whether this is normal outburst behavior or what its long-term behavior really is. More observations are needed to understand this system.

Figure 1.10. The UGSU-type dwarf nova V478 Her fading to magnitude 16.5 after an outburst. Imaged on April 26, 2007, by the author with a Celestron 14 SCT + SBIG ST9XE @ f/7.7. 120-second exposure. North is at the top.

V358 Lyr
18 h 59 m 32.95s +42° 24′ 12.2″

This enigmatic CV is seriously under-observed, not least because its maximum magnitude is 16 and it slumps to 20 or fainter at minimum. In the Information Bulletin on Variable Stars (IBVS 5544,1) in 2004, Antipin, Samus, and Kroll argued that the maximum magnitude of this alleged nova was the result of a 1965 photographic plate defect, but it still remains a mysterious object with the official classification uncertain between it being a nova and a WZ Sge-type dwarf nova.

V336
Per 03 h 22 m 53.84s +41° 37′ 01.4″

This star is another enigmatic object at a high northerly declination. Professional astronomers Liu and Hu, writing in the *Astrophysical Journal Supplement Series* (128:387-401, 2000 May) confirmed its spectroscopic identity as a CV in November 1998, but it is rarely seen by amateur observers. The U.K.'s Gary Poyner suspected it at magnitude 15.9 on January 11, 2004, although an image this author secured the next night showed it at minimum. Nevertheless, as a confirmed CV with an outburst magnitude as high as 14.3, it is a high-priority object.

TY Vul
20 h 41m 44.0s +25° 35′ 11.0″

At minimum this star sits as low as magnitude 19, close to the limit of short CCD exposures with modest-aperture telescopes. But in outburst it can reach as high as magnitude 14. Regular monitoring shows that it experiences low-level outbursts where it fluctuates between magnitudes 16.6 and 16.0. This is too faint for all but the most determined (and experienced) visual observers, even with large telescopes, but it is ideal for amateur CCD imagers. TY Vul experienced a rare outburst at the start of December 2003 when it was spotted by Patrick Schmeer at magnitude 14.9. The previous outburst was in September 1999, when it reached magnitude 14.6. During the 2003 outburst Dave Messier (Connecticut, U.S.) was the first to detect superhumps in TY Vul's light-curve, thus defining it as a UGSU-type dwarf nova. Analysis of the data by Tonny Vanmunster revealed a periodicity of just less than 2 hours. But despite this success TY Vul has rarely been seen in outburst, so continuous monitoring and more outburst observations are required.

Leading CV Observers

Gary Poyner

Without doubt the world's greatest visual CV observer must be the U.K.'s Gary Poyner (see Figure 1.11). The sheer number of visual observations he makes every year is staggering in itself (around 15,000 in a good year, mainly CVs!!) but

Figure 1.11. The world's most prolific CV observer, Gary Poyner of Birmingham, Great Britain. Poyner has made over 200,000 visual magnitude estimates since 1975 and can reach magnitude 16.8 visually with his 14″ (356 mm) LX200. Image: Jean and Gary Poyner.

incomprehensibly impressive when you take account of the fact that he lives in a mainly cloudy country and observes from the light-polluted city of Birmingham! Residents of that city have a distinctive accent and are known as 'Brummies,' and Gary signs all his e-mails with the line: 'You can always tell a Brummie....but you can't tell him much'!

Gary started off as a planetary observer, but at age 17 (1975) he became fascinated by the appearance of the brilliant Nova Cygni of that year in the August sky and became hooked on variable star observing, especially CVs. Over the years Gary has used a variety of telescopes, but the largest have been 40-cm and 45-cm Dobsonians and a 35-cm Go To Schmidt–Cassegrain. With all these telescopes he has been able to observe and estimate the magnitudes of up to 150 stars in a night! How can this be possible? Even with a Go To instrument most people would become pretty tired after slewing to, and casually viewing, 150 objects. But moving the telescope manually and accurately estimating the magnitudes of them, too? Surely such a feat is superhuman? In fact, the human brain is quite a remarkable instrument, and, over millions of years of evolution (hundreds of millions for the more basic instinctive functions) has made the recognition of patterns, faces, friends, and enemies something that is built into all humans. Even today the world's most powerful pattern recognition software cannot match a human being when it comes to recognizing human faces from all angles and when disguised. It turns out that this human ability can perfectly adapt itself to looking at star fields, too.

Britain's George Alcock, mentioned elsewhere in this book, memorized 30,000 stars in patterns through his binoculars, and the Australian supernova hunter Bob Evans memorized more than 1000 galaxy fields. Gary Poyner has simply memorized hundreds of faint star fields, along with their comparison star magnitudes and where they reside in the sky. So, maybe he is not superhuman after all. But he is bordering on superhuman!! In 2007 Gary passed the 200,000 variable star estimates milestone. In recent years, when it is cloudy, Gary has started using the Bradford Robotic Telescope (based in Tenerife) for observing faint CVs over the Internet. In Gary's own words:

"I don't think I'm superhuman! When you've observed a variable star field a few hundred times you just remember where to push the scope, and you remember the comparison stars, too. So you don't need a chart or a torch, you just instinctively hop from field to field. I've memorized the fields of about 400 variables now. It just becomes second nature. It takes about one minute to find and estimate each star's magnitude, so you can do 150 stars in about 150 minutes. I remember observing in December 1982 when it was –23 C! The eyepiece was so cold I got a skin burn on my eye socket after observing for an hour! As long as my head and feet are well covered I manage to cope. Plus, we don't seem to get cold winters anymore. I got the LX200 to make observing more comfortable. To be honest, I can find the fields quicker by pushing a Dob around by hand, but then you have to clamber up to the eyepiece. I'm also having to re-learn the 400 fields that are mirror-imaged in the LX200 diagonal. I had a fright the other night. I slewed to the field of a new variable in Hercules, observed it, then realized with horror that I didn't actually know where it was in the sky. It was a weird feeling just typing in the RA and Dec. The LX200 optics are very good, though: the coatings and baffling at f/10 seem to give a very dark field at high powers. I've seen down to mag 16.8 with it, a shade fainter than with the larger 16 and 18″ Dobs. I don't think I have exceptional eyesight. I tend to wear dark glasses in the daytime to protect my eyes from dazzling sunlight, but I think seeing almost to mag 17 just comes from experience. Your eyes and brain just learn to see fainter stars. My wife Jean has known I was a mad astronomer from the first time we met. She is incredibly supportive of my hobby and comes to look through the eyepiece if anything special happens. I'd have to resort to stamp collecting if not for Jean's support. I have only annoyed her once. I was sick in bed with a bad case of tonsillitis, when a phone call came through that UV Per was in outburst and that confirmation was needed. I got up, took my portable 22 cm outside, and observed the outburst through a gap in the cloud with heavy snow falling all around and on me (which is why I didn't open the observatory to use the 16 inch), and feeling very ill, too. Still I did get confirmation, my 22 cm mirror got a good washing, and Jean eventually forgave me (after about a week!)"

What does Gary judge to be the highlights of his observing career?

"I guess there are many highlights, but most are probably important to me alone—like my first 15th & 16th mag stars, or the faintest mag limit reached, early/confirmatory observations of novae etc. There have been a number of 'firsts,' like picking up several outbursts of stars never seen before (such as V1113 Cyg and a number of more obscure objects), and the first ever fade detected in the (then suspected) VY Scl star V751 Cyg, which confirmed its type. Margareta Westlund and I were also instrumental in helping a new class of CV to be discovered—the UGSU-ER UMa class, through our very early and intense monitoring of this system. It's always nice to play a role like that. DV UMa is important to me, as I was on the phone to Tonny Vanmunster in Belgium the night he detected the first ever outburst. We were both looking at it through our respective eyepieces and chatting, separated by hundreds of miles, and we were the first people ever to see it. The following outburst I observed the first ever eclipse in this system, and the one following that I sat through an entire orbital period to get consecutive eclipses. This meant staring through the eyepiece for more than 2 hours, making an estimate every 60 seconds. I also made the first ever visual observations of an eclipse in the dwarf novae EX Dra. In those days there were no CCD's, so I was the first amateur to do that.

I was monitoring V635 Cas (an X-ray novalike system containing a giant star and a pulsar) for outbursts at the same time as Diane Dupre (at Los Alamos) was monitoring in the X-ray, and we had a good liaison going with that one. I observed a particular outburst in the late 90's and told her about it. She was amazed as the X-ray flare hadn't been seen (they usually precede any optical brightening). She picked it up the next night, which was the first time it had outbursted in the optical before X-ray. She was very happy if I can remember correctly.

There was also one occasion when Guy Hurst rang me to ask if I could confirm whether IR Gem was in outburst, as he had an astronomer asking for confirmation of a dodgy outburst report, and was waiting to turn EUVE onto it, but didn't want to if it wasn't in outburst. I had to quickly go outside to check the weather, to find a fortuitous gap in the cloud and the full Moon extremely close to IR Gem! Pressure was on, as Guy was waiting for me to call back. I confirmed the outburst (which was tough with the Moon oh so very close); EUVE made some UV obs, and everyone was happy.

But the most memorable event of all must be the flickering I saw in 1RXSJ053234.9+624755 (also known as 'Bernhard 01 Cam'). This was in April 2005, and the star was just reported to be in outburst. I looked at it and couldn't believe what I was seeing. It was varying by as much as one magnitude in just tens of seconds. After the initial shock, I monitored it for 24 minutes, making an estimate every 10 seconds. My eye nearly fell out after 24 mins, so I gave up as anything done after that would be poor data. I've has seen flickering before in lots of CV's, but this was the most intense. Jeremy Shears and I wrote a paper about the outburst (*Journal of the British Astronomical Association* Vol. 116, No. 1). I still marvels at the processes which can make a star or disk do that! Jeremy got some excellent photometry on it, too (see Figure 1.12).

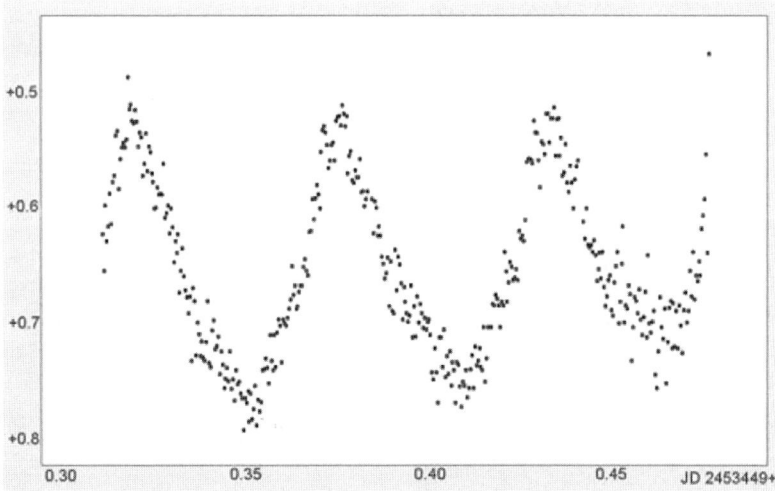

Figure 1.12. Remarkable superhumps in 1RXS J053234.9+624755 on March 19, 2005 from 19.30 to 23.39 UT. Magnitudes relative to a magnitude-11.2 comparison star. Photometry by Jeremy Shears with a 100-mm Takahashi refractor and Starlight Express MX716 CCD.

Jeremy Shears

Jeremy lives at Bunbury, Cheshire, in England and has established an impressive reputation for detecting rare outbursts of some of the less well-studied dwarf novae (see Figure 1.13). In addition, he has carried out very high-quality time-series photometry for many outbursts, the results of which have been published in the *Journal of the British Astronomical Association*. Up to 2006 Jeremy's main instrument was a relatively small-aperture (102 mm) Takahashi FS102 with a focal length of 820 mm. Using this equipment on a Vixen GPDX mount (Skysensor SS2K Go To), and a Starlight Xpress MX716 CCD camera, Jeremy typically patrolled 50–70 CV fields in each observing session, looking for outbursts. The wide field of the Takahashi/MX716 combination worked well with the Go To accuracy of the Vixen mount. However, recently Jeremy has upgraded his equipment to a Celestron 11 optical tube on a Gemini G41 computer-controlled mount (with a Boxsdorfer DynoStar X3 hand controller). The equipment is housed in a '7 foot' (2.1 m) Pulsar Dome. Although he could reach magnitude 18 in 60-second exposures with the Takahashi refractor, using the 280-mm Celestron means better photometry is possible on the fainter stars.

Jeremy looks for the much sought-after superhump evidence in his time-resolved photometry as well as evidence for eclipses. The CVs that Jeremy routinely patrols are generally those on the BAA VSS/TA ROP list, maintained by Gary Poyner, but also those discovered by the Hamburg Quasar survey (Boris Gaensicke's list). In addition, new CVs discovered by the SDSS have become top priority objects in recent years. To quote Jeremy: "It's quite a thrill to be the first human to spot such an outburst, and then there's the anticipation of finding out what type of CV it is and whether it's the highly prized UGSU-type of dwarf nova."

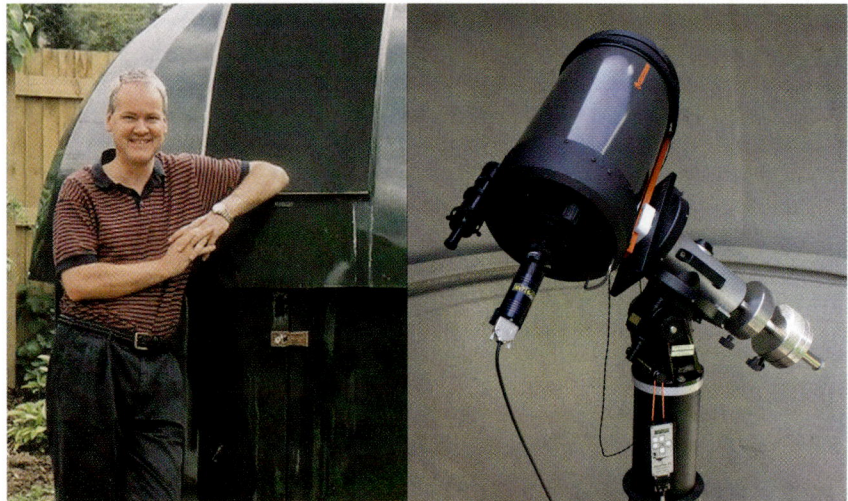

Figure 1.13. Jeremy Shears of Bunbury Observatory, Cheshire, specializes in spotting dwarf novae as they go into outburst. Originally his dome housed a 100-mm Takahashi refractor, but a Celestron 11 (280 mm) on a Gemini G41 mount is now used for more sensitive photometry. Images: Jeremy Shears.

Mike Simonsen

Mike Simonsen, shown in Figure 1.14 lives near Imlay City, Michigan, and is one of the world's keenest CV observers. He is also the administrator of CVnet and hosts their webpages. His observing location is in a rural area and so enjoys fairly dark skies. On his website Mike describes himself as a variable star junkie, namely:

> "A person who is willing to brave the elements and sleep deprivation, while investing substantial sums of money, time, and spousal permission units to observe stars that, for a variety of reasons, fluctuate in brightness on timescales of minutes, hours, days, or years. Once addicted, the VS junkie can never get enough telescope time and spends much of his day fretting over the weather, planning observing sessions, analyzing data, organizing variable star charts, surfing the web, and reading books on variable star related topics."

Mike has named his two 12-inch (30 cm) LX200 Schmidt–Cassegrains after friends, namely, Arne Henden of the AAVSO and the late and much missed

Figure 1.14. Michigan amateur Mike Simonsen is one of the world's leading CV observers. Simonsen is shown here in his new (2007) roll-off roof observatory housing his second 12-inch (30 cm) LX200 Schmidt–Cassegrain, named Arne. His older 12-inch Schmidt–Cassegrain, named Janet, is housed in the dome in the background. Image by kind permission of Mike and Irene Simonsen.

rise, and announcing it in time for spectra to be obtained during the rise, is of great value and highly prestigious. Most nova spectra are obtained at or after maximum.

It is thought that classical novae recur every 10,000 years or so, although, for obvious reasons, this is yet to be proven. Although classical nova outbursts are some 11-magnitudes fainter, in absolute terms, than typical supernova explosions, they are still incredibly bright in normal stellar terms. A typical, classical nova situated closer to us than 1000 light-years will have a negative magnitude. If one of our nearest stellar neighbors, such as Sirius, went nova, it would shine as brightly as the full Moon for the days or weeks after the outburst. Rather obviously, most novae are thousands, or tens of thousands, of light-years away.

In the twentieth century one nova stood head and shoulders above the rest, namely V603 Aquilae, discovered by G.N. Bower (and many others!) on the night of June 8/9, 1918. It rapidly peaked at magnitude –1.1, but in only 8 days had dropped down to magnitude 2. It is still easily visible in a modest telescope as a star drifting in brightness between magnitudes 11.3 and 12.0. Coincidentally a total solar eclipse had tracked across the United States in the hours before this nova erupted. The famous comet discoverer and Ohio amateur astronomer Leslier Peltier mentioned this stunning nova in his biography *Starlight Nights*. He was carrying his portable refractor and mount to a position where he could observe variable stars in the southern sky when he saw Nova Aquilae 1918 blazing down:

"That was the night that I forgot all about telescopes and variables for as I turned and looked up at the sky, right there in front of me – squarely in the center of the Milky Way – was a bright and blazing star! I have always wished that I could recapture my sensations of those first few minutes of that sighting. That I was bewildered and confused goes without saying, for I had acquired a fair knowledge of the stars and constellations and here, right before me, was a total stranger, a star that had not been there just the night before."

Of course, in 1918, the lights of planes heading toward the observer were never going to pose as novae! Plates exposed at Harvard Observatory on the previous night (June 7, 1918) showed that famous nova just bordering on naked-eye visibility at magnitude 6.

Novae of easy and obvious naked-eye visibility (i.e., magnitude 4 and brighter) are rare, but even so, a few will occur in every observer's lifetime. A photograph of 4th magnitude Nova Cygni 1992 is shown in Figure 2.1.

Apart from the obvious attraction of novae as observing targets, that is, their spectacular appearance in the sky and the brightness of the best examples, they are especially fascinating because no two nova light-curves are the same. This is because, among other things, the following parameters are variables:

- The white dwarf mass, temperature, and C, N, O content
- The secondary mass and mass transfer rate
- The separation/rotation periods of the two components
- The presence or absence of eclipses, which are, fairly obviously, line-of-sight dependent
- Accretion disk stability and mass-transfer fluctuations
- The absence/presence of strong magnetic fields in the system.

In the study of novae the term t_3 is often seen. This is the time it takes for a nova to drop 3 magnitudes from its peak magnitude. In general, a fast nova (NA) is defined as one for which t_3 is less than 100 days (often *much* less); a slow nova (NB)

Figure 2.1. The bright Nova Cygni 1992 which peaked on the night of February 21, when this photograph was taken. The star was at magnitude 4.3 that night. 0.36-m f/5 Newtonian, 19:03–19:19 UT Hypersensitized Kodak 2415. Field 1-degree high with north up. Image by the author.

is defined as one for which t_3 is greater than or equal to 100 days. There is also a third category (NC), defined as being a *very* slow nova and taking a decade or more to fade from maximum. Obviously a nova is far more observable when it fades slowly; for example, in the case of George Alcock's discovery, HR Del, t_3 was determined to be an impressive 230 days. There will be more about Alcock later in this chapter.

A large number of bright, slow-fading novae have been well studied over the last 40 years by amateur astronomers. Some of the most comprehensively studied examples are described a little later in this chapter.

Associated Nebulosity

Only 6 months after the shy and modest Scotsman Thomas Anderson discovered the nova GK Per in 1901, French astronomers, led by Camille Flammarion, announced what they described as 'a luminous shell' surrounding the star. Of course, the natural instinct would have been to assume that material had been flung outward from the recent nova eruption and that was what was being observed. However, you do not need to be a genius to work out that, at a distance of 1500 light-years, even nebulosity extending an arcminute from the star, after 6 months, must imply an ejection speed close to the speed of light. Indeed, after astronomers observed the shell's expansion rate they initially concluded that knots of material were moving at speeds well in excess of the speed of light.

Of course, an explanation for this apparent violation of the laws of physics was soon found. When GK Per's eruption was observed here on Earth, the actual explosion had taken place 1500 years earlier. On its way to us the light, spreading out in all directions from the nova, had hit a sheet of dust in front of the nova. The light-hitting regions of this sheet, well-removed from the central (nova) position as viewed from Earth, would only take fractionally longer to arrive than the light nearer to the nova. Thus, it looks like faster-than-light travel is involved. It is like slamming a disk several light-years in diameter onto a sheet of glass between us and the nova. The edges of the disk arrive at the same time as the middle, making it look as if there was instant travel from disk center to disk edge.

How odd that there was a sheet of dust there, you might say? Well, not really, as many novae and supernovae appear to puff out gas and dust before erupting, and even classical novae probably recur on timescales of thousands or tens of thousands of years. (A famous 'light echo' was photographed around Supernova 1987A when that exploded.) The environment within light-months and light-years of novae is often very dusty. The nebulosity around GK Per is still visible today in long-exposure H-alpha images as a faint, arcminute-diameter, spoked-wheel shape often referred to as the 'Firework nebula.'

When a nova explosion occurs material is hurled into space at several thousand kilometers per second and quickly envelopes the secondary star region, whether light-seconds or light-months away. The orbital motion of the two stars serves to stir the gas and dust up, ensuring intricate patterns in the shells when viewed at high resolution. As recently as 2007 a faint wisp of nebulosity was observed on deep H-alpha images at the position of Nova Vulpeculae 2007, just before it flared to binocular visibility. When the nova was discovered on August 8 by the Japanese patroller Hiroshi Abe, H-alpha plates taken by the Isaac Newton Telescope on June 26 were examined and revealed obvious nebulosity in the region of the progenitor star on those images, taken some 7 weeks earlier. In 2003 NASA's Galaxy Evolution Explorer, imaging in the ultraviolet, detected a massive shell around the star, indicating that it may have experienced a colossal nova-like outburst thousands of years ago.

Memorable Novae

There have been many memorable novae over the years, but I have selected a few personal favorites in this section. Some are novae that are easily recalled; others are ones that are simply memorable for historical reasons. This is especially applicable to the novae discovered by the English school teacher George Alcock.

Table 2.1. The brightest (magnitude 3.5 and above) novae of the twentieth century

Designation	Discovery year	Peak magnitude	Discoverer
GK Per	1901	0.0	Anderson
DN Gem	1912	3.3	Enebo
V603 Aql	1918	−1.1	Bower
V476 Cyg	1920	2.0	Denning
RR Pictoris[1]	1925	1.1	Watson
DQ Her[2]	1934	1.2	Prentice
CP Lac[3]	1936	1.9	Gomi et al.
CP Puppis	1942	0.4	Dawson
V553 Her	1963	3.2	Dahlgren and Peltier
HR Del	1967	3.5	Alcock
V1500 Cyg	1975	1.8	Osada
V382 Vel	1999	2.6	Williams/Gilmore

[1]RR Pic took 150 days to fade to magnitude 4.1.
[2]DQ Her took 94 days to drop to magnitude 4.2.
[3]CP Lac was also independently discovered by Nielsen and Loreta. Nielsen was on a cruise ship at the time of the discovery: the P&O lines SS Strathaird. The cruise had been partly organized by the BAA comet photographer Reggie Waterfield, to allow BAA members to view the June 19, 1936, total solar eclipse from the island of Chios in the Aegaen Sea. One of Nielsen's fellow travelers on the cruise was the stage and screen comedian Will Hay.

HR Del
20 h 42m 20.30s +19° 09′ 39.0″

George Alcock swept up Nova Delphinus on July 8, 1967; it was, and is, the only nova to have been discovered in Delphinus. HR Del was the first British nova to be discovered since J. P. Manning-Prentice's equally magnificent DQ Her of 1934. Prentice was Alcock's mentor, but the apprentice went on to far exceed the master's achievements! The nova rose quickly to 4th magnitude and an extended pre-maximum halt; it then rose again to a peak of magnitude 3.5 on December 13, 1967. The nova then dropped in brightness, rising again to another peak of magnitude 4.2 on May 5, 1968. A more leisurely decline to 11th magnitude by 1974 (and 12th, eventually) then took place. An analysis of the BAA VSS observations of HR Del from 1967 to 1971 reveals a total of 3657 BAA visual observations, with the top observer being none other than Sir Patrick Moore, who submitted 436 estimates!

LV Vul
19 h 48 m 0.70s +27° 10′ 20.0″

George Alcock's second nova discovery, 9 months after the first, was a much faster nova, but HR Del was still bright when No. 2 flared up, leading to a dramatic sight in the morning sky. The nova LV Vul was swept up by George on April 14, 1968, rising to a peak magnitude of 4.8 a week later, on April 21. With HR Del on the rise to its final 4th-magnitude peak, there were two British naked-eye novae, only 15 degrees apart, in the April dawn sky. For a full account see the excellent biography of George Alcock, *Under an English Heaven*, by Kay Williams.

V1500 Cyg
21 h 11m 36.60s +48° 09′ 02.0″

On August 29, 1975, the brightest nova since 1934 (Prentice's DQ Her) erupted in Cygnus. V1500 Cyg peaked at magnitude 1.8 on August 31, but with a t_3 of only 3.6 days, it faded rapidly. A week after outburst it had totally faded from view. BAA member Ken Kennedy secured some of the best photographs of the time (see Figure 2.2). Those who waited for a really clear night, or for the next weekend, were very disappointed! Hundreds of amateurs worldwide claimed Nova Cyg 1975 as their own; Cygnus was briefly and unmistakably different to anyone who had even the slightest knowledge of constellations. The official discoverer is sometimes given as Minoru Honda; indeed, he was one of the first. But the very first person to file a discovery claim was another Japanese observer, Osada.

With a magnitude 21.5 progenitor, the amplitude range of V1500 Cyg was truly remarkable. In the United States it was a long holiday weekend, resulting in the Central Bureau for Astronomical Telegrams' (CBAT) not issuing a circular until September 2, long after various worldwide telephone calls had been made. The IAU Circular (2826) listed the first discoverers as Osada, Honda, Ito, 30 members of Nihon University Astronomical Study Group, and Hashimoto, along with dozens of other non-Japanese discoverers across the world.

Figure 2.2. Nova Cygni 1975, now designated V1500 Cyg, completely changed the appearance of Cygnus, but it faded rapidly, too. Ken Kennedy captured some excellent pictures of the nova at its peak. The image here was taken by Kennedy on September 1, 1975, with an Exakta camera and 200-mm telephoto lens. The film was Kodak Spectroscopic 103a-f emulsion, and a sodium filter (ON16) was used to cut the street light glare. Exposure time was 7 minutes guided. The field has north at the top and is roughly 7 degrees wide. The nova is arrowed, and the bright star Deneb, slightly distorted, appears in the lower right. Image by kind permission of Ken Kennedy.

V838 Her
18 h 46 m 31.50s +12° 14′ 02.0″

Some 8 years after his final (fifth) comet discovery George Alcock swept up his fifth and final nova in the dawn sky on the morning of March 25, 1991. This was a remarkable discovery in many ways. Firstly, Alcock had a strong feeling that he was going to be lucky that night, so strong in fact that he was not at all surprised when he spotted the 5th-magnitude intruder; with the Milky Way memorized down to magnitude 8 or 9, an object this bright was unmistakable. Secondly, as with Comet Iras-Araki-Alcock in 1983, Alcock discovered the new object from indoors, through a double-glazed window, in this case using nothing more than hand-held 10×50 binoculars. Thirdly, the confirmation itself was equally remarkable. At the time of discovery, 0435 UT, nautical twilight had arrived! Alcock knew from his weather satellite images (he had a big dish in his garden well before the Internet era) that Conder Brow observatory near Lancaster was likely to be clear, so he gave Denis Buczynski a phone call. Within minutes of being woken, a semiclad (time was short) Denis was in the dome, and the astrograph was being slewed to the nova. At the time of the first exposure the sky was already so bright that only Deneb was visible!

Despite this, Denis secured a photographic plate and within the hour had developed and fixed it and identified the new object. It is sometimes questionable as to whether telephoning observers to confirm a discovery in the early hours is really necessary; however, it was crucial in this case. Nova Her was at its peak when Alcock spotted it, and it dropped rapidly in magnitude ($t_3 = 2.8$ days) in the following days and weeks. Denis and Guy Hurst were quick to alert the astronomical community, and, from Coonabarabran, New South Wales, Rob McNaught also secured a photograph of the nova at its peak; only 12 hours after the discovery it was already down to magnitude 6.5. Professional astronomers were quick to study the rapidly fading nova, and subsequent research indicated that fluctuations in the accretion disk appeared remarkably quickly after the nova peaked (due to the rapid fade of the primary photosphere).

V1974 Cyg
20 h 30 m 31.70s +52° 37′ 51.0″

Less than a year after Nova Herculis, on February 19, 1992, the best-studied nova in history was discovered visually by Peter Collins. Nova Cyg 1992 (V1974 Cyg) rose to a peak of magnitude 4.4 by February 21. (The strongest memory this author has of any nova is my recollection of guiding an exposure for 20 minutes on that evening, with Nova Cyg as the guide star and then viewing the nova through my 0.36-m Newtonian. The star had a distinctly dazzling golden appearance through the telescope; it just looked special. Minutes after I closed the shutter the field had sunk below the roof of the house next door; Cygnus is badly placed in evening and morning skies in February, even from England.)

V1974 Cyg was comprehensively studied because so many key astronomical facilities were operating at the time, including ROSAT, IUE, HST, KAO, Compton GRO, KAO, and Voyager. The decline of V1974 Cyg was reasonably leisurely and uneventful, apart from a few plateaus in the decline ($t3 = 38$ days), but in 1996, with the nova just above magnitude 16, the decline virtually stopped, and the nova has been around magnitude 15.6 to 15.8 ever since. Guy Hurst has drawn attention to this feature of the nova a number of times and this author has taken a number of confirmatory images of the field over the last few years. When the nova was first discovered, astrometry by Skiff (Lowell Observatory) and comparison with the Palomar Sky Survey suggested an 18th -magnitude progenitor. The belief that the progenitor was 18th magnitude seems to have survived to this day, despite known problems with the original astrometry, but further investigation into this led to the conclusion that the nova is some 2.5" away from the 18th -magnitude star; astrometry by Denis Buczynski suggested this as early as 1992. This is clearly shown in an H-alpha image of the nebulosity surrounding the nova, taken by the WIYN telescope and featured in the October 1996 edition of Sky & Telescope.

V1419 Aql
19 h 13 m 6.80s + 01 ° 34 ′ 23.0 ″

Novae are often neglected when faint or when the initial period of enthusiasm has faded; the same is true of supernovae. Guy Hurst, of the British magazine *The Astronomer* (TA), has frequently appealed to observers to be vigilant and follow objects down the light-curve as far as possible. It is remarkable how many novae and supernovae decide to do something abnormal just as observers are becoming bored with them! This was especially true in the case of V1419 Aquilae (Nova Aquila 1993). This magnitude-7.6 object was discovered on May 14 by Yamamoto using a 200-mm lens. The nova faded quickly, but without drama, to 14th magnitude by the end of July. With little happening and the star approaching many observers' visual limit, the observations tailed off. Rather predictably, the star then started to brighten again and reached magnitude 12.5 by December. Unfortunately, solar conjunction then rendered the field unobservable until April 1994, but once again the message was clear: never relax your vigilance on a fading nova. In February 1995 Takamizawa discovered another nova in Aquila, V1425 Aql, but that one was far better behaved.

V705 Cas
23 h 41m 47.20s + 57 ° 31 ′ 01.0 ″

Two of the most extraordinary novae of recent years were discovered in a northern Milky Way constellation that had previously yielded no confirmed classical novae (although two possible nova-like variables in Cassiopeia are listed by Hilmar Duerbeck in his book *A Reference Catalogue and Atlas of Galactic Nova'*). The 1993 and 1995 novae in Cassiopeia have been studied intensely by amateur astronomers, and the resultant light-curves are comprehensive and remarkable. By comparison, the 1993 and 1995 novae in Aquila were positively boring!

Nova Cass 1993 (V705 Cas) was discovered at magnitude 6.5 by Kanatsu on December 7, 1993, using a 55-mm lens at f/2.8 and T-Max 400 film. It reached maximum light on December 21. V705 Cas, as it became known, took 2 months to fade by 2.5 magnitudes (see Figure 2.3). But suddenly, around February 14, 1994, it went into a dramatic decline, confirming what some professionals had been expecting; this was a rare, dusty, DQ Her-type nova, and the dust, ejected in the nova outburst, was now rapidly cutting off the light from the system. The nova dropped from magnitude 9 to magnitude 12.2 between February 14 and February

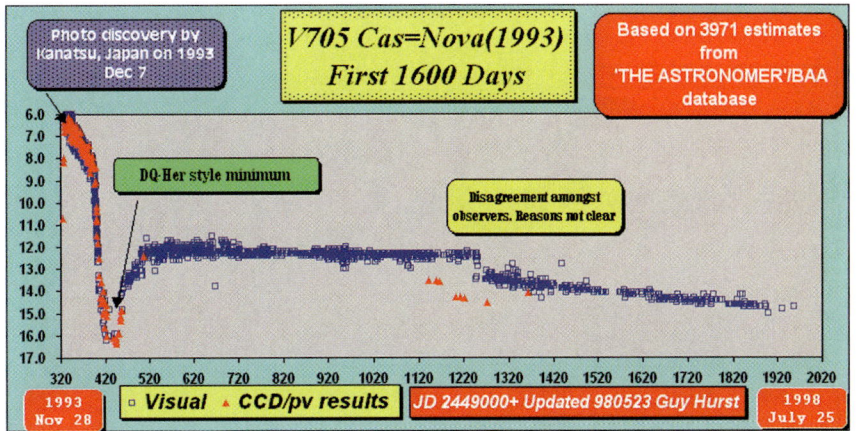

Figure 2.3. The light-curve of V705 Cas = Nova Cas 1993, discovered by the Japanese patroller Kanatsu on Dec. 7, 1993. The nova entered a deep 'DQ Her'-type fade when dust from the outburst obscured the star. But as the dust dissipated it brightened again and stayed at 12th magnitude for the next 2 years! Light-curve by kind permission of Guy Hurst/The Astronomer.

20. The questions were how long would the dust take to clear, and how deep would the decline be? Most visual observers lost the star at around 15th magnitude; fortunately, Cassiopeia is a circumpolar constellation, so light-curve data was not confused by solar conjunction.

At this point photographic and CCD observers took over, but caution was required here. Although most CCD images look very similar to the photographic field, CCDs are sensitive in the infrared; red stars will look brighter in CCD images than they do in the visual or V-band. Light passing through dust is reddened as the longer wavelengths tend to get through more easily; hence, V705 Cas, in its DQ Her fade stage, appeared noticeably brighter in unfiltered CCD images than in the V-band or in photographs. V705 Cas bottomed out at just below 16th magnitude. The keenest visual observers did not lose it for long, and a few months after its dramatic drop from 9th magnitude to 16th magnitude, it had risen back to be a 12th-magnitude object. But the fun was not over yet. In the summer of 1996, 2.5 years after the outburst, the nova suddenly dropped by a magnitude from its comfortable plateau, and a slow decline commenced. There was considerable disagreement among visual observers at this time as to the magnitude of the star. In addition, CCD images showed the star to be fainter than the visual estimates, by up to 2 magnitudes. It seems likely that color anomalies were to blame, with the nova, with the comparison stars, and possibly with both. The matter has never been fully resolved, but it proves that nova light-curves are far from predictable or explicable!

V723 Cas
01 h 05 m 05.40s + 54 ° 00 ′ 40.0 ″

The second proven nova in Cassiopeia was discovered by Yamamoto on August 24, 1995, at magnitude 9.2; the discovery was made with a 200-mm f/4 lens. At first this just seemed to be a rather faint and not too interesting nova; however, pre-discovery images by Kosaka (IAUC 6214) from July 30 to August 23 indicated

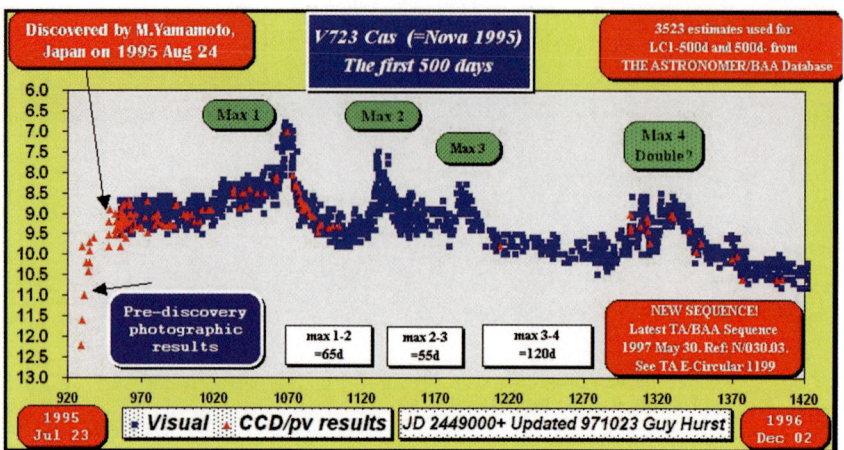

Figure 2.4. The truly bizarre light-curve of V723 Cas = Nova Cas 1995, discovered by the Japanese patroller Yamamoto on August 24, 1995. The nova originally peaked almost 5 months after its discovery and then exhibited another three peaks in brightness in the next 10 months. Light-curve by kind permission of Guy Hurst/The Astronomer.

the object had risen slowly from magnitude 12.2 to magnitude 9.2 over this period; it had not suddenly burst into view, like most novae. After discovery, V723 Cas rose slowly in brightness for a while but then suddenly accelerated to a peak of magnitude 7.1 on December 16/17. By December 22 it was back down to magnitude 8.4. The slump continued to magnitude 9.5 some 40 days after the peak, and then another peak occurred in mid-Feb 1996, with apparently rapid fluctuations in brightness on a timescale of hours. A third, 8th-magnitude maximum occurred 55 days later and a fourth 120 days after that! A fifth (weak 10th magnitude) peak was also observed in August 1996. The light-curve shown in Figure 2.4 is somewhat reminiscent of HR Del, except that nova was 3 or 4 magnitudes brighter. The jury is still out on the precise reason for this extraordinary behavior.

V2362 Cyg
21 h 11m 32.30s + 44 ° 48 ′ 04.0 ″

This nova (Figure 2.5) was discovered as a new magnitude 10.5 object on April 2, 2006, and, once again, the discoverer was a dedicated Japanese patroller, Hideo Nishimura. The progenitor star was discovered on archival images and was just below magnitude 20. Within 2 days of discovery the nova had peaked at 8th magnitude and was on the downward slide. Eight weeks after discovery it had faded back down to 12th magnitude, but then the decline stopped, and it sat at that magnitude for over 3 months. A steady rise then ensued, and V2362 Cyg gradually crept back up in brightness. Eight months after its outburst the nova was back at magnitude 10, fully 10 magnitudes above its quiescent state (see Figure 2.6). The performance was a bit reminiscent of the 1999 nova V1493 Aql, but rather slower. Theories abound as to the cause of this bizarre behavior. Maybe this re-brightening was simply due to a shell of dust clearing? Maybe a second nova

Figure 2.5. Nova Cygni 2006 (V2362 Cyg) imaged some 6 months after its discovery by the Japanese patroller Nishimura. Image by the author, with a 35-cm Celestron 14 at f/7.7 and an SBIG ST9XE CCD on Oct. 13, 2006. This was a 60-second exposure, and the field is 13 arcminutes wide with north at the top.

outburst had taken place, or maybe we had witnessed a nova outburst followed by a dwarf nova outburst? The jury is still out on this recent nova, too. By the end of 2006 V2362 Cyg had faded to 13th magnitude and a very nearby star, almost touching the nova on amateur images, was complicating the magnitude assessment.

V5558 Sagittarii
18 h 10 m 18.27s −18 ° 46 ′ 52.1 ″

This nova was discovered on April 14, 2007, by Yukio Sakurai, on two 20-second exposures with a Fuji FinePix S2 Pro Digital SLR and a 180-mm focal length f/2.8 Nikon lens. The nova was at magnitude 10.3 on the discovery images. Two months later, in mid-June, the nova was still creeping upward in brightness and had reached magnitude 8.5. It looked as if the rise was flattening out, and despite its rather unusual and leisurely increase astronomers expected a gradual decline to follow. They were wrong. Suddenly, it was off again, and rising at a steeper rate than before! By July 8 it had reached magnitude 7.0, an easy binocular object. It then started fluctuating randomly between about magnitude 8.8 and magnitude 7.5, in some ways reminiscent of V723 Cas some 12 years earlier – another one for theorists to puzzle over. Sadly, its declination of almost −19° made it a very challenging object for far northern hemisphere observers, but those further south covered V5558 Sagittarii's behavior on every available night.

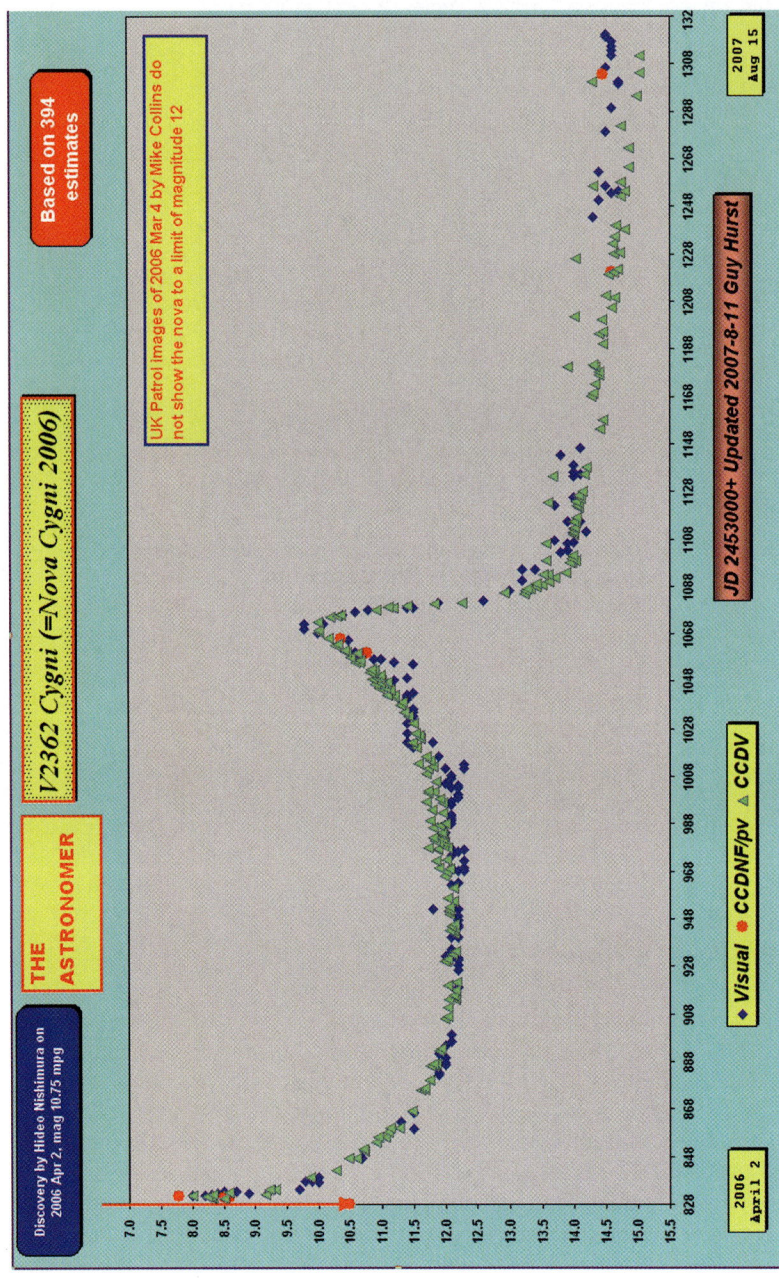

Figure 2.6. The highly unusual light-curve of Nova Cygni 2006 (see previous figure for the star field), which peaked at 8th magnitude, fell rapidly to 12th magnitude, and then rose again to brighter than 10th magnitude before suffering a second dramatic fall. Light-curve by kind permission of Guy Hurst/The Astronomer.

Recurrent Novae

In recurrent novae (NR), the frequency of recurrence is typically 10–80 years. Not surprisingly, proven examples of these objects have to exist for this subdivision to have been created, but they are exceptionally rare, with only nine recurrent novae known in our galaxy (see Table 2.8 in the next section for full details). Perhaps the best examples are RS Ophiuchi, with its six outbursts, and T Corona Borealis, which can reach magnitude 2.

It might be thought that an outburst interval greater than 10 years would easily exclude confusion with dwarf novae, but this is not necessarily the case. For example, the dwarf nova HV Vir was originally classified as a classical nova when it outburst in 1929. A second outburst was observed in 1929, when follow-up photometry and spectroscopy showed it to actually be a rare WZ Sge-type dwarf nova.

Recurrent novae have outbursts of about 4–9 magnitudes. It would appear that many of these systems contain a giant secondary star and a large mass transfer rate, hence the frequency of the outbursts. The spectroscopic properties of recurrent novae, and their light-curves, suggest these are fully fledged novae, as opposed to dwarf novae, but the decline from maximum light tends to be rather faster than for the normal nonrecurrent (classical) novae. We have to be careful here, though, because the implied opposite (i.e., nonrecurrent novae) cannot mean that the classical novae themselves never recur. Astronomers have only been scouring the Milky Way systematically and efficiently during the last century, so many recurrences of classical novae may have occurred in centuries gone by. The recurrent novae may simply be novae that we know have had more than one outburst in the last century, that is, in the modern era, when the sky is patrolled vigorously.

Professional astronomers are not entirely clueless as to what is happening in these systems, though. Although there are many unique examples that test the theories and lead to different papers being submitted to learned bodies by competing university departments, there is a general consensus as to what leads to the recurrence of nova outbursts. Systems containing white dwarfs of a substantial mass are thought to have faster nova eruptions. In addition, the massive white dwarf novae have smaller radii and a much greater gravitational field; thus, they do not need to accrete as much material as a lighter white dwarf to trigger a nova explosion.

Only 30 millionths of a solar mass (i.e., 10 Earth masses of hydrogen) needs to accrete onto a heavy 1.3-solar-mass white dwarf to trigger a nova. Far more material has to accrete onto a lighter white dwarf of about half the mass, say, 0.6 solar masses; in this lighter case, you need roughly 5000 millionths of a solar mass (1670 Earth masses) to reach the trigger point. If we assume an accretion transfer rate from secondary star to white dwarf of around a three-thousandth of an Earth mass per year, then it will take 30,000 years for the 1.3-solar-mass white dwarf to get to nova detonation point, and 5 million years for the lighter 0.6-solar-mass white dwarf. Put simply, as you get nearer the white dwarf 1.4-solar-mass 'Chandrasekhar' limit (the Type Ia supernova point), the accretion time to nova outburst drops rapidly. You would therefore expect all high-mass white dwarf systems to recur at faster and faster rates as their mass increases. Mind you, we have left one vital part of the equation out here, namely, how much material is lost in each nova explosion: is it more, or less, than was accreted? The general view is that most white dwarfs going nova may lose slightly more mass than they accrete in a nova outburst; thus, they are doomed to a longer period between outbursts. Nevertheless, in, say, a billion-year lifespan, that still leaves plenty of time for loads of outbursts. But the heaviest-mass white dwarfs may be an exception to this shrinking trend; that is, there is little to stop them gaining mass and eventually evolving to reach that Type Ia supernova/1.4-solar-mass point. Once they get there and really go bang, they will certainly not recur again. But they will have ended their lives with a hell of a bang! See Chapter 4 for more details. One thing is for sure – the few recurrent novae that have outbursts every decade or so are available for intense scrutiny every time they outburst, and thus the theories can be revised in the lifetime of an astronomer.

But not all astronomers are happy that recurrent novae simply slot conveniently into this simple 'higher mass equals more frequent outbursts' model. What if the recurrents are not simply higher-mass white dwarf versions of classical novae? Ronald Webbink and colleagues published a paper in Volume 314 of the *Astrophysical Journal* in 1987 ("The Nature of Recurrent Novae") where they suggested different criteria to qualify an eruptive star as a recurrent nova. As well as stipulating an absolute visual magnitude of less than or equal to –5.5 and a shell ejection velocity of greater than or equal to 300 km per second, the same team proposed that there might actually be two classifications of recurrent novae, based on the fact that their outburst intervals are too frequent to be classical novae and too rare to be dwarf novae. In other words, rather than being unusual variants of classical novae, recurrent novae may sometimes be nearer in type to dwarf novae.

Webbink et al. proposed that recurrent novae of Type A, like the star T Pyxis, were indeed classical novae; that is, they were the result of a thermonuclear runaway on the surface of a white dwarf. Conversely, Type B eruptions, in their theory, are more like dwarf novae; that is, they are caused by accretion processes from a red giant onto the companion's disk. In such events an outburst may be generated by instability in the cool companion or by instabilities in the disk. In this classification system the recurrent novae T CrB, RS Oph, and V745 Sco would classify as Type B. Confused? No surprise there! Unfortunately, in astronomy, CV stars rarely fit precisely into well-defined boxes, and there are always going to be objects that will have astronomers debating whether they are novae, recurrent novae, or dwarf novae. The spectroscopic and photometric data have to be analyzed and weighed up with the recurrence interval so that it is clear whether a thermonuclear explosion has taken place on the surface of a white dwarf, or whether an accretion disk is the source of the outburst mechanism. When enough data has been collected, and the binary star modeled, the object can, hopefully, be sorted into a classical, recurrent, or dwarf nova box. However, this will rarely stop professional astronomers debating the definition of borderline objects when they only outburst every few decades. Whichever way you look at it, monitoring recurrent novae, and even unusual classical novae, for outbursts, is valuable work that all amateurs with appropriate equipment can get involved with.

How to Observe Novae and Recurrent Novae

Novae often fade very quickly from view, and even with CCDs, it is rare for anything really dramatic to still be happening more than a few months after the initial outburst. So the work that an amateur can do tends to fall into three distinct categories: namely, searching for new novae and the fame that goes with a discovery; checking old and faint recurrent novae in case they flare up; observing brand-new novae intensively as soon as they are discovered and are still bright. There is, arguably, a fourth category, too, but one that is a very specialist niche – obtaining spectra of the brighter novae. Although you need a large, or very large, amateur telescope to capture spectra of a bright supernova, a nova spectra gives you far more light to play with.

Visual Patrolling

Most amateur astronomers will have heard of the five comets and five novae discovered visually by George Alcock (1912–2000) (Figure 2.7) from the back garden of his Farcet home, near Peterborough in England. Alcock's nova discovery technique, which has since been copied by a few others, was unique when he first achieved success in the 1960s: simply memorize all the stars easily visible in the northern Milky Way through hand-held 15 × 80 binoculars! He also independently discovered an outburst of the recurrent nova RS Ophiuchi in 1985 while patrolling for novae; it had first been spotted, a few days earlier, by

Figure 2.7. The remarkable English schoolteacher George Alcock (1912–2000), pictured here around 1988, (aged 76 or so) with his 25 × 105 tripod-mounted binoculars and (hung from his shoulders) his 15 × 80 Beck-Tordalk binoculars. Alcock is looking out of an upstairs window (in the room where he kept his radio and weather satellite equipment), which gave him a clear view over the flat Peterborough fenland. By this time he had discovered five comets and four novae, as well as an independent discovery of an outburst of RS Ophiuchi. His fifth nova discovery (Nova Herculis 1991) would come 3 years later. Alcock committed 30,000 stars, in patterns, to memory. Image: BAA Archives.

W. Morrison of Ontario. Coincidentally, both men lived close to towns named Peterborough in their respective countries! It has been estimated that Alcock could recognize approximately 30,000 stars, in patterns, using his technique. In the 1960s and 1970s he was the only man able to consistently steal novae away from the Japanese. Alcock discovered his five novae in 1967, 1968, 1970, 1976, and 1991, as well as independently discovering the RS Ophiuchi recurrent nova outburst in 1985.

It is easy to assume that Alcock must have had a photographic memory to achieve this feat, but others have copied his methods, including the U.S. nova discoverer Peter Collins and the Portugese patroller Alfredo Pereira, with similar results, although from much clearer skies. Pereira uses 9 × 34 and 14 × 100 binoculars and has memorized several thousand stars in a total of 14 Milky Way constellations. Regions are studied down to magnitudes between 7.5 and 9.0, depending on the constellation. Pereira tends to use the larger binoculars for sweeping the lower -altitude regions, such as Sagittarius, saving the lighter 9x 34's for the more neck-kricking higher -altitude regions. Like Alcock, some of Pereira's sweeping is carried out from indoors, while looking out of a window. After his 1999 Nova Aquila discovery in December of that year, Pereira stated he had spent about 100 hours per year sweeping the skies in 1998 and 1999, with a grand total of 500 hours of sweeping completed prior to that discovery. Pereira has now discovered or co-discovered four novae visually namely: V1494 Aquilae, V4739 Sgr, V4740 Sgr, and V597 Puppis. These were captured in December 1999, August 2001, September 2001, and November 2007. His most recent nova in Puppis was discovered after a total of 626 hours of binocular sweeping in the six 6 years since September 2001, from his site at Carnaxide in Portugal.

Peter Collins of Tucson, Arizona, preceded Pereira in emulating George Alcock's visual nova-hunting techniques, although with the distinct advantage of living in Arizona. Collins has stated that he tends to use "the smallest binoculars that will reach 8th magnitude comfortably under the prevailing conditions of illumination and atmosphere," with 7 × 50s showing too many stars from his favorite 2500-meter mountain site at Mt. Hopkins in Arizona. He also observes from the 'Burnham Lookout' point of the Steward Observatory and the roof of the observatory building in downtown Tucson. Occasionally he uses much heavier 11 × 80 binoculars, too. Collins has discovered or co-discovered four novae visually: namely, V1668 Cyg, QU Vul, QV Vul, and V1974 Cyg. These were caught in September 1978, December 1984, November 1987, and February 1992. The May 2001 nova in Aquila (V1548 Aql) was discovered by another Collins, namely the British photographic patroller, Mike Collins.

Many variable star observers have found they can easily memorize the star fields of more than 100 subjects, and the Australian supernova discoverer, Bob Evans, has memorized more than 1000 galaxy fields. The human brain is exceptionally good at pattern recognition; think of how many human faces you can identify: your friends, film stars, politicians, criminals, newsreaders. We can all recognize hundreds or even thousands of faces, without even knowing we are doing it! The recognition process all occurs in a tiny section of the brain, too. When that part is damaged in, for example, a road accident, the unfortunate person can lose all ability to recognize anyone, including their own face in the mirror! Yet other parts of the brain can be totally unscathed. Although a photographic memory may not be essential for nova patrolling, patience, determination, and sensitive night vision are extremely important.

So what made the legendary George Alcock decide to discover novae visually? Well, he had been primarily searching for comets and made his fourth discovery in September 1965 (he would not discover his fifth comet for another 18 years). The 6-year period from 1959 to 1965 marked Alcock's golden period of comet discovery, but even in the 1960s he was wondering whether, with increasing light pollution from Peterborough, he could discover any more. From his visual meteor work of the past 20 years, Alcock could recognize about a thousand stars in patterns. This does not mean he could name them all, or that he could even draw the constellations down to magnitude 5 or so. It simply meant that his brain's pattern recognition center could tell if a new star was breaking up the old pattern. He started to wonder if he could take this into another league and memorize the prime nova search fields to magnitude 8 or so. Even though it was his idea, the thought of committing the Milky Way, as seen through binoculars, to memory was "preposterous" even to Alcock; it implied memorizing perhaps 20,000 or 30,000 stars. But taking the easy way out had never been an option for Alcock. He would search for novae, but continue looking out for comets, too.

From 1967 to 1976 Alcock used his memory of the northern Milky Way to full advantage, sweeping up four novae in a 11-year period. He had rivals in this field, too, most notably the Japanese photographic patrollers; Honda was a major rival. But Alcock was the only successful observer at that time searching visually. On clear nights he had a huge advantage over the photographers. Bright novae would be spotted almost instantly by him; there was no additional hassle of developing films, mounting negatives, and stereomerging or blinking. He could also observe in an instant and between cloud banks. His visual approach had much more flexibility than photography. His first nova success came on July 8, 1967, when he swept up Nova Delphini rising through 6th magnitude. At last, the 12 years of memorizing the Milky Way through binoculars had paid off; it must have been a huge relief. Alcock now had five discoveries to his credit: four comets and a nova.

As we saw earlier, in the late seventies and eighties, the American observer Peter Collins would follow Alcock's example and memorize the Milky Way stars, too, but Alcock had shown the way: Nova Delphini 1967, or HR Del as it came to be known, is still the only nova to have been discovered in Delphinus. It was the first British nova to be discovered since Prentice discovered DQ Her in 1934. The nova rose to a peak of 3.5 on December 13, 1967, dropped slowly, then peaked again at 4.2 on May 5, 1968 – an extraordinary object. Alcock's second nova was discovered a mere 9 months after the first and was a much faster nova. This one was in Vulpecula and was designated LV Vul. It was discovered on April 14, 1968, rising to a peak of magnitude 4.8 a week later, on April 21. Remarkably, and as noted earlier but worth repeating, with HR Del on the rise to its final 4th-magnitude peak, there were two British naked-eye novae, only 15 degrees apart, in the April 1968 dawn sky! Alcock has often stated that the sight of those two novae together was "the greatest thrill of my observing career." The proximity of LV Vul to the bright nova of 1670 was also of considerable importance to Alcock. Two years later, Alcock notched up his third and faintest (at magnitude 6.9) nova, in Scutum, V368 Scuti. Another 6 years would elapse before he bagged his fourth, on October 21, 1976, NQ Vul, a nova right next to the famous Coathanger asterism. This was an especially important discovery for Alcock, as his morale was somewhat dented by 'missing' the 1st-magnitude nova V1500 Cyg on August 29, 1975, the day after his 63rd birthday. This would have given him a really bright nova to beat that of his mentor, Prentice. Alcock only missed the spectacular nova by a few hours, and the loss nearly made him give up. Alcock described his feelings toward Nova Cyg 1975 at a meeting of the JAS (now SPA) on April 29, 1978, at the Holborn Library in London.

His final discovery, on March 25, 1991, when Alcock was 78 years old, was a remarkable one in many ways, and we have already described the twilight confirmation of V838 Her by Denis Buczynski in the memorable novae section. An independent discovery of this final Alcock discovery was made by Sugano in Japan, and V838 Her was one of the fastest fading novae of all time, dropping 3 magnitudes in 2.8 days. As an 18-year-old Alcock had been told by the 90-year-old meteor observer Grace Cook that she thought he would take the place of W. F. Denning, who discovered five comets and a nova from England. In many ways, Alcock replaced Denning and the Scotsman T. D. Anderson, who, in terms of personality (quiet, modest, and somewhat reclusive) was a lot more like Alcock then Denning was. Denning died in 1931 and Anderson one year later.

Other British Nova Discoverers

British observers have a fine tradition of discovering novae visually. Hind, Anderson, Espin, Denning, and Prentice all discovered bright novae between 1848 and 1934. Huddersfield postman John Hosty (Figure 2.8) managed to snatch one nova

Figure 2.8. Huddersfield postman John Hosty (1949–2001), the discoverer of Nova Sagittae 1977, with his 50 mm monocular (half of a pair of binoculars). He discovered Nova Sagittae (HS Sge) on January 7, 1977, as a star of magnitude 7.2, low down in the western winter early evening sky. Only 3 weeks later it had faded below magnitude 10. Image: The Astronomer Archives.

from Alcock in 1977, and the Scottish astronomer Rob McNaught discovered three southern hemisphere novae, with basic amateur photographic equipment, from Siding Spring, in the late 1980s. Of course, he has now gone on to be a major discoverer of all types of objects, and there are now more comets named after McNaught than after any other human being. John Hosty described his discovery of the nova HS Sge on January 7, 1977, in the February 1977 issue of *The Astronomer* magazine:

"I decided this particular evening to make an estimate of a suspect variable near IQ Per. From 17.15–17.30 UT I swept the region which includes α Per and a 10° area was cleared to mag 6. At 17.30 I suddenly decided to sweep Aquila-Sagitta (British nova patrol area 75). After 5 minutes sweeping from Altair to α Sge, I suddenly noticed a star close to α which I could not recognise from memory on previous sweeps of this region. I immediately became quite excited as I realised that this could possibly be a nova. After the initial surprise and amazement I noted the position and made a field sketch, estimating the magnitude to be 0.3 brighter than a star immediately below α Sge."

In 1988, Midlands amateur Dave McAdam discovered a nova-like variable in Andromeda on his photographs (PQ And, now thought to be a dwarf nova); that object was confirmed by myself and Guy Hurst. But, in recent years only Mike Collins has discovered a genuine nova from English soil: Nova Aquilae 2001. This was also a photographic discovery, made after 13 years of patrolling and after Mike had discovered 200 other variable stars (Table 2.2).

In April 1968, Alcock's first two novae were both visible to the naked eye, in the same part of the dawn sky! (Tables 2.3 and 2.4)

Table 2.2. Classical galactic novae (T Cor.B was later found to recur in 1946) discovered from the UK and Ireland arranged by date of birth order of their discoverers. This list does not include Robert McNaught's novae (discovered from Australia) or Dave McAdam's dwarf nova PQ And, discovered in 1988. If the enigmatic object UW Per, discovered by C. R. D'Esterre in 1912, was a classical nova, then it should be included

Discoverer	Lifespan	UK location	Equipment	Novae	Constellations
J. Birmingham	1814–1884	Galway (Ire)	Naked eye	1	T Corona Borealis (1866)
J.R. Hind	1823–1895	London	18 cm refr.	1	V841 Oph (1848)
W.F. Denning[1]	1848–1931	Bristol	Naked eye	1	V476 Cyg (1920)
T. D. Anderson[2]	1853–1932	Edinburgh	Binocs/Eye	2	T Aur (1891/2); GK Per (1901)
T.H.E.C. Espin	1858–1934	Tow Law	Astrograph	1	DI Lac (1910)
J.P.M. Prentice	1903–1981	Stowmarket	Naked eye	1	DQ Her (1934)
G.E.D. Alcock	1912–2000	Farcet	Binoculars	5 + 1	See Table 2.3
J. Hosty	1949–2001	Huddersfield	Monocular	1	HS Sge (1977)
M. Collins	Born 1950	Sandy, Beds	135 mm lens	1	V1548 Aql (2001)

[1]Denning independently discovered Nova Aql 1918 (V603 Aql), but after the official discovery by Bower on a Harvard Observatory photographic plate. It was a hard object to miss!

[2]Thomas Anderson was such a shy and modest observer that when he discovered T Aurigae in Feb. 1892 he announced it by sending an anonymous postcard to Professor Ralph Copeland at the Royal Observatory, Edinburgh. It transpired (by checking Harvard Observatory plates) that the nova had initially flared up 2 months earlier, in mid-December 1891. He spotted his second nova, GK Per, when walking back to his Edinburgh home on Feb 21, 1901. He imagined it had been visible for some time and so was somewhat surprised when the first nova of the new century was confirmed as his discovery.

Table 2.3. The five novae discoveries and one independent recurrent nova discovery made by George Alcock

Alcock Discovery (Recovery for RS Oph)	Date	Independent discovery
HR Del (= Nova Del 1967)	July 8th 1967	None
LV Vul (= Nova Vul 1968 No 1)	April 15th 1968	None
V368 Sct (= Nova Sct 1970)	July 31st 1970	None
NQ Vul (= Nova Vul 1976)	October 21st 1976	None
RS Oph's 5th outburst	January 30th 1985	Morrison (Canada)
V838 Her (= Nova Her 1991)	March 25th 1991	Sugano (Japan)

Table 2.4. Novae of magnitude 6.0 or brighter from 1967 to 2007. Note the four discoveries and one independent recovery by George Alcock. Alcock's only nova fainter than magnitude 6.0 was Nova Scuti 1970 (V368 Scuti), discovered at magnitude 6.9

Designation	Discovery year	Peak magnitude	Discoverer
HR Del	1967	3.5	Alcock
LV Vul	1968	4.8	Alcock
FH Ser	1970	4.4	Honda
V1500 Cyg	1975	1.8	Osada
NQ Vul	1976	6.0	Alcock
V1370 Aql	1982	6.0	Honda
QU Vul	1984	5.6	Collins
RS Oph (recurrent)	1985	5.4	Morrison/Alcock
V842 Cen	1986	4.6	McNaught
V838 Her	1991	5.0	Alcock
V1974 Cyg	1992	4.3	Collins
V382 Vel	1999	2.6	Williams/Gilmore
V1494 Aql	1999	3.8	Pereira
RS Oph (recurrent)	2006	4.5	Narumi/Kanai
V1280 Sco	2007	3.7	Nakamura/Sakurai

DSLR Patrolling

The sane alternative to visual nova patrolling used to be photography, which, in the film-dominated era up to the end of the twentieth century, meant using black and white emulsion that was developed very quickly and then checked visually using a stereo viewer (with a master negative in one eyepiece view) or a blink comparator. One of the strongest advocates of photographic nova patrolling was the American amateur Ben Mayer who, in 1978, invented the term PROBLICOM, standing for 'Projection Blink Comparator.' Two slide projectors, which could be aligned to project their images onto the same spot, were used to project a master negative and a recent patrol image onto a screen. A rotating propeller mounted between the two projectors chopped the images up so that one image appeared, followed by another. Thus any nova would blink; ingenious, but all rather fiddly and slow.

Since the DSLR era dawned, nova patrolling has become far less messy and less fraught, because DSLR images can be developed without noxious chemicals and they can be digitally blink compared fairly quickly. Unlike with supernova patrolling you only need cover half a dozen wide field regions, not thousands of galaxy fields, and you do not need to spend tens of thousands of dollars on equipment, either. However, as soon as technology becomes affordable by individuals and by university departments there will always be competition. Various 'wide field' CCD-based detectors scan the night sky automatically from dark and clear locations each night, and several novae have already been detected by some of these professional patrols. Nevertheless, amateurs equipped with standard DSLRs (and photographic SLRs), especially Japanese amateurs and Bill Liller (based in Chile), continue to dominate the field of nova discovery. Interestingly, in 2007, the Australian amateur Terry Lovejoy discovered two comets with a 'typical' nova patrol set up, that is, a DSLR plus a 200-mm lens. In passing we might mention that many people must have been drooling when, in 2007, Canon announced it was about to release a brand-new 200-mm lens with the staggering f-ratio of 2; in other words, a 200-mm lens with a 100-mm aperture! If you have deep pockets and want to shorten your nova patrol exposures, that might be the lens to choose.

The longer your lens focal length, the less area you will capture, of course, but the longer lens will go deeper. A camera lens with a usable aperture of at least 40 mms is a useful size to aim at and almost guarantees a limiting magnitude of 11 in all but the most light-polluted skies. Thus, an 85-mm lens at f/2, or a 135-mm lens at f/2.8 is a good weapon. A fast, high-quality lens, such as, a 55-mm f/1.4 model, will go deep and deliver at least a 24-degree wide field (even with the two-thirds-sized chip detectors), but exposures longer than 1 minute will be saturated by skyglow. With even a modest 135-mm f/2.8 lens novae as faint as magnitude 10 can be discovered; they are far enough above the image limit to register as a definite suspect. At the time of writing, affordable DSLR detectors are rapidly moving back toward the size of the old '35-mm' film frames, 36×24 mm. The earlier reference to two-thirds-size detectors refers to the smaller CMOS and CCD detectors used in the early or budget DSLRs, such as Canon's Digital Rebel/300D camera. The imaging area of these camera's chips is two-thirds the size of an old 35-mm film frame, roughly 15×23 mm. Sometimes this size is referred to as APS (Advanced Photographic System) format; although, strictly speaking, APS is a film format with three different sizes.

Where to Search

Ask any experienced northern hemisphere nova patrollers where to start patrolling and the chances are they will advise you to start with Ophiuchus, Aquila, Cygnus, or Vulpecula (see Figure 2.9). A relatively narrow constellation to the south of Cygnus, Vulpecula has had more than its fair share of novae, even for a constellation in the Milky Way. At the south-western edge of Vulpecula lies the distinctive asterism called The Coathanger, a memorable pattern of stars that is an ideal place for the visual patroller to start developing those pattern-recognition skills. A remarkable number of nova photographs contain this asterism in the field. The whole region around the famous Coathanger asterism has a history of 5th- and 6th-magnitude nova discoveries, and one discovered as long ago as 1670 (CK Vul) reached 2nd magnitude.

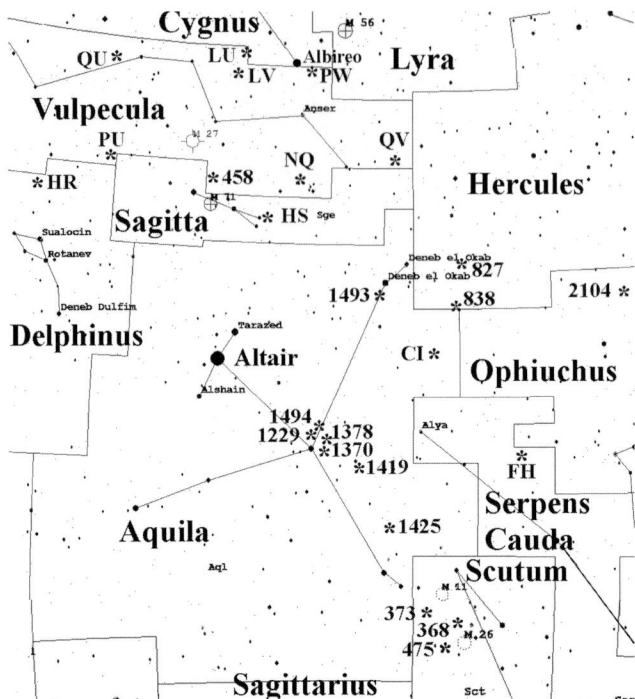

Figure 2.9. This diagram shows 25 novae (marked with an asterisk) of magnitude 10.0 or brighter that were discovered between 1967 and 2007, between declination –10 and +30 and roughly 18–21 h of Right Ascension, that is, from southern Cygnus to southern Aquila. The letter/number designations refer to the constellation prefix of each nova; that is, HS = HS Sge and 458 refers to V458 Vul. NQ Vul is very close to the 'Coathanger' asterism.

Table 2.5 shows the total number of novae discovered in the 12 most productive constellations up to the end of 2007. Of course, some of these constellations are much bigger than others, and the most productive nova regions span constellation borders (as one would expect!). Sagittarius (Figure 2.10) is the most productive nova constellation and the second most productive per unit area, too, although many novae near the galactic center are only magnitude 9 or 10. Tiny Scutum is the most productive constellation on a 'novae per square degree basis. The second most productive constellation is Scorpius.

Of course, the northerly parts of Sagittarius and Scorpius are virtually unpatrollable from the far northern hemisphere, even in their summer months. The most productive constellation that is easily accessible (at least in summer and autumn) from the northern hemisphere is Aquila, with Cygnus and Vulpecula prime targets, too. Scutum, Serpens Cauda, and Ophiuchus should not be ignored by far northern hemisphere patrollers, but it all depends how low you can get in those summer and autumn months. In practice, when you look at a star chart and examine where all the Aquilae novae tend to occur, you realize that there is a sort of fertile rectangle not far from Altair, which is well worth aiming your camera at; this rectangle has opposing corners at 20 hours R.A., –10 Dec. and 18 hours R.A., +30 Dec. The middle of the rectangle's eastern edge would just contain the bright star Altair – a very useful marker for aiming your camera.

Table 2.5. All-time total nova discoveries in the most productive constellations up to the end of 2007. Compiled from IAU records

Constellation	Novae	Comments
Sagittarius	93	Galactic Center. S. Hem. Masses of faint novae.
Scorpius	35	Galactic Center. S. Hem. Masses of faint novae.
Ophiuchus	33	Prime region, accessible to both hemispheres.
Aquila	28	Prime region, accessible to both hemispheres.
Cygnus	17	Prime N. Hemisphere region, esp for bright novae
Scutum	16	Visible from both hemispheres. Tiny constellation.
Centaurus	11	S. Hem
Vulpecula	10	Especially productive near the 'Coathanger'
Hercules	8	Eastern edge worth monitoring (DQ Her, V553 Her)
Serpens (Both parts)	6 + 2	2-part constellation. 6 novae in Cauda; 2 in Caput
Puppis	8	S. Hem
Carina	7	Far south, productive Milky Way region

With a full format 36 × 24 mm DSLR sensor, aligned with the longest sensor axis north–south, a lens of 50-mm focal length would almost cover this rectangle. Alternatively, two frames, one above the other, with the sensor aligned east–west and a 70-mm focal length, would also do the job.

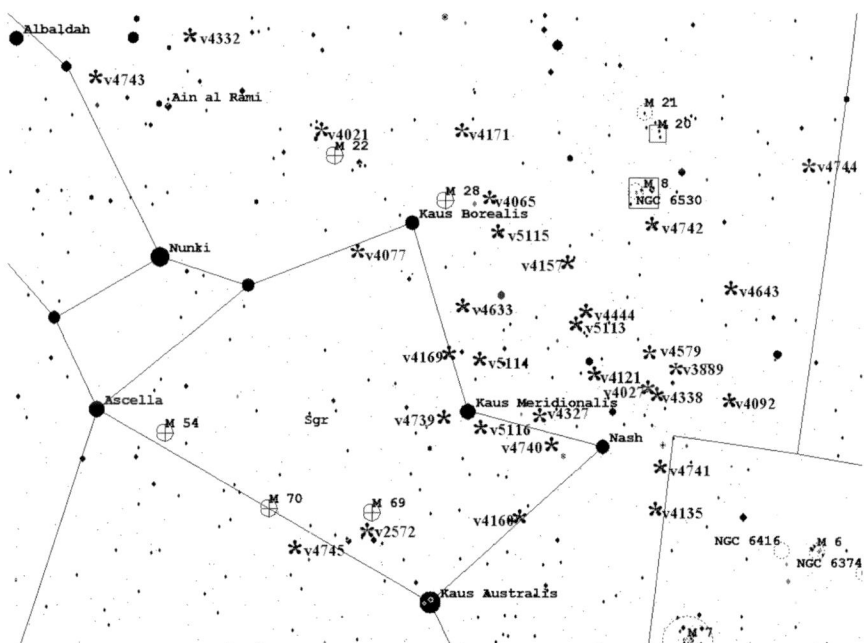

Figure 2.10. This diagram shows 31 novae in central Sagittarius of magnitude 10.5 or brighter that were discovered between 1968 and 2007. Note how many novae there are near the 'spout' of the teapot. The V number designations indicate the variable name in Sagittarius, for example, V4021 Sgr.

Moving north through Aquila, the Aquila/Hercules border is a fairly productive region, and then we come to the fertile nova spawning grounds of Sagitta, Vulpecula, and southern Cygnus, near Albireo. Moving further north takes us into central and northern Cygnus. This far away from the galactic center, novae are a bit sparser but often quite spectacular, as a lot of relatively nearby novae have occurred here. The bright novae of 1876, 1920, 1975, and 1992 were all within 10 degrees of Deneb. As we move further north, the number drops further, but those that have been found have been bright and unusual. The northern Cygnus/Lacerta/Cepheus border regions have produced a dozen novae since the late nineteenth century. In addition, the two novae discovered in Cassiopeia, in 1993 and 1995, were some of the most unusual ones ever observed, and the 1901 Nova, GK Persei, reached 0 magnitude.

For northern European and North American nova patrollers, Ophiuchus and Scutum are only well-placed around the summer months. The worlds leading nova discoverer, Bill Liller, of Vina del Mar, Chile, has discovered numerous novae in the galactic center, but from his southerly latitude he does have a distinct advantage. With the ecliptic passing through this region, extreme care has to be exercised, since bright asteroids can easily look like novae. The nova patroller soon finds that he or she not only needs decent star atlases showing all faint variable stars but a good planetarium package, like *Guide 8.0*, too.

Despite the fact that the circumpolar regions above 40 degrees north are accessible all year round to British observers, they are neglected hunting grounds. As mentioned earlier, in the 1990s two extraordinary novae were discovered in Cassiopeia, and yet they are the only *proven* novae in this constellation; one is forced to the conclusion that many novae in this far northern constellation have been missed. (We mentioned earlier that Duerbeck's Atlas, listed in the Resources section of this book, mentions two possible faint historical novae in Cassiopeia, BC Cas and V630 Cas, but neither are categorically proven as novae and both are below 10th magnitude).

The winter Milky Way constellations have also produced a few novae, but again, looking out for asteroids is very important. Gemini has produced four novae in total, but none recently; again, evidence indicates that the constellation is either being neglected or deliberately avoided because of its proximity to the ecliptic plane (Table 2.6).

Table 2.6. The density of novae per constellation, measured in units of all-time nova discoveries per 100 square degrees. The figures can be somewhat misleading if taken at face value, as some constellations have a concentration of novae in a small area, with few elsewhere

Con.	SCU	SGR	SCO	AQL	OPH	VUL	SER[1]	CYG	CAR	PUP	CEN	HER
Density	14.7	10.7	7.0	4.3	3.5	3.4	2.9	2.1	1.4	1.2	1.0	0.7

[1] The value for Serpens is calculated just for Serpens Cauda, the easterly tail of the Serpent. Confusingly, Serpens is split into two, the other half (westerly) being the head (Serpens Caput). Serpens has enjoyed a total of eight novae, six in the Milky Way Cauda half and two bordering the Milky Way in the Caput half. The overall value for Serpens works out at 1.3 novae per 100 square degrees. In reality the Milky Way stretching from western Aquila, down into Scutum, Sagittarius, and Scorpio is the most densely packed nova hunting ground, with Aquila and Scutum being the only practical patrol regions in that selection for many northern hemisphere nova hunters. Tiny Scutum has the highest nova density per unit area.

Successful Photographic Nova Hunters and their Equipment

Minoru Honda and his Acolytes

The Japanese have a formidable reputation in nova hunting largely because of the 14 (photographic) discoveries of Minoru Honda (1913–1990). He inspired a whole new generation of photographic nova and comet hunters, namely Takamizawa, Kuwano, Wakuda, Sugano, Sakurai, Suzuki, Yamamoto, Tago, Haseda, Nishimura, Nakamura, Takao, and Hatayama. Pictures of Minoru Honda and some of those Japanese photographic patrollers he inspired are shown in Figures 2.11–2.18. Even today, with automated patrol systems such as ASAS (All Sky Automated Survey with a 180-mm-focal-length f/2.8 lens) detecting a few novae, the Japanese patrollers and Bill Liller (based at Vina del Mar, Chile) still seem to just have the edge in claiming the discovery. The Japanese did not have a reputation as nova patrollers until Honda discovered FH Ser on February 13, 1970. That single discovery, by an already legendary comet discoverer, would have an astounding effect for the next four decades at least. Honda's further 13 nova discoveries in the following 17 years inspired his countrymen Kuwano and Wakuda to successes themselves, with Suzuki and Sugano joining the club prior to Honda's final discovery. In the

Figure 2.11. The legendary Japanese nova and comet discoverer Minoru Honda (1913–1990) poses at his "Ask to stars" observatory, with its forest of telescopes, binoculars, and astrographs on December 7, 1987. Honda discovered 12 comets visually between 1940 and 1968 and then, like Alcock, switched to nova hunting and discovered 14 novae photographically (three were co-discoveries) between 1970 and 1987. His novae were discovered in Serpens (2), Vulpecula, Cygnus (2), Crater, Sagittarius (3), Aquila (3), Lacerta, and Hercules. Photograph by kind permission of Tomohisa Ohno, Kurashiki, Japan. Relayed to the author via Osamu Ohshima and Seiichiro Kiyota.

Figure 2.12. The prolific Japanese nova discoverer Yukio Sakurai of Mito, Ibaraki-ken, with some of his astronomical equipment. (A 180-mm-focal-length f/2.8 lens plus Digital SLR has been used for his discoveries.) Sakurai has discovered (5) or co-discovered (2) a total of seven novae since 1994. Five of these finds were in Sagittarius, one in Scorpius, and one in Puppis. Image by kind permission of Yukio Sakurai, communicated to the author by Seiichiro Kiyota.

37 years since Honda's very first nova discovery, he and the countrymen he inspired would discover or co-discover 59 of the 157 Milky Way novae between FH Ser (1970) and V458 Vul (2007). That equates to a Japanese nova discovery/co-discovery every 8 months over that 37-year period.

Bill Liller

Of course, amateur Japanese nova hunters are particularly highly motivated individuals who enjoy 'beating' the professionals to a discovery. But Bill Liller (Figure 2.19) is not Japanese and has been a professional astronomer himself; he is a retired Harvard University professor who moved to Chile on his retirement. Situated there, at Vina del Mar, he is right under the Sagittarius Milky Way center and far better placed than the Japanese to patrol the heart of our galaxy. This combination of determination, location, and retired status has enabled him to discover a staggering number of objects. Liller has discovered over 40 CVs of various types, that is, classical, recurrent, dwarf, and extragalactic (LMC/SMC) novae as well as a comet (1988 V), two supernovae, and asteroid 3040 Kozai. Most of Liller's novae have been discovered with an 85-mm f/1.4 Nikon lens and Kodak Tech Pan 2415 film. He also uses an orange filter to increase contrast and because novae have a reddish H-alpha tint to them. Many visual observers have reported bright novae as looking 'gold' to the eye.

For a number of years in the 1980s and 1990s Rob McNaught (1986–1987) and Paul Camilleri (1991–1993) gave Liller some stiff southern hemisphere competition, but these days his main competition is from the few Japanese patrollers who

Figure 2.13. Yuji Nakamura of Kameyama, Mie, Japan, shown here with his Takahashi EM-200 mount and Pentax camera. Nakamura discovered six novae between 2001 and 2007 using a 135-mm-focal-length lens. Two of these were co-discoveries. The novae were discovered in Scorpius (2), Ophiuchus (2), Sagittarius, and Cygnus. Image by kind permission of Yuji Nakamura, communicated to the author by Seiichiro Kiyota.

can scour the sky with the same ruthless efficiency. Liller is on record as saying: "Chilean women are dangerous! I met an attractive widow in 1979, couldn't resist her charms, and moved in with her in 1981." Taking retirement from his Harvard University post in his early fifties (he was born in 1927), and moving to a latitude of 33 degrees south, were the vital steps in Liller becoming the king of the nova patrollers, even outstripping the discoveries of the Japanese. His discoveries started a year after his patrols, when he found a new 7th-magnitude star in the far southern constellation of Muscae, at a declination of –67. Up to the end of 2007 he had been credited with the discovery or co-discovery of 38 novae within our Milky Way. Thirteen of these have been in Sagittarius, with six each in Norma and Centaurus. The rest have been found in Scorpio (4), Circinus (3), Muscae (2) and 1 each in the constellations of Aquila, Crux Australis, Fornax, and Lupus. In recent years Liller has used an old 20-cm-aperture f/1.5 Celestron Schmidt camera, plus CCD, to obtain spectra of his discoveries and to speed up the official confirmation process.

Figure 2.16. By day, Akira Takao of Kita-kyushu, Japan, is a doctor of medicine, but by night he is yet another successful Japanese nova patroller. Takao discovered four novae (three were co-discoveries) between 2003 and 2005. His current nova patrol system may well be unique, especially in Japan, where most patrollers use Takahashi equatorial mountings! Takao uses the fork from an old LX200 Schmidt-Cassegrain with a wooden box slung between the fork tines holding a 120-mm-focal-length f/3.5 lens attached to an SBIG ST8 CCD. This gives a field of view of 6.6 × 4.4 degrees. Image by kind permission of Akira Takao, communicated to the author by Seiichiro Kiyota.

also required for exposures of more than a few seconds duration. Some nova patrollers simply piggyback their patrol cameras on existing telescope mounts, whereas others buy small equatorial mounts such as the Takahashi Sky Patrol 2. Some even make motorized 'barn door' mounts (a hinge forms the polar axis between two plates, and a motorized threaded rod opens the door). Recently (2007) the highly portable but accurate AstroTrac TT320 has shown that a system does not have to be bulky to produce good tracking over many minutes, even with long lenses.

The Japanese nova patroller Kesao Takamizawa, shown with a pair of giant binoculars in Figure 2.17, has been a keen observer of comets, variable stars, novae, and supernovae since the 1960s. He started photographic patrol work in 1994, and the next 5 years were his most productive. His main instrument during that period was a twin 10-cm-aperture 400-mm-focal-length lens system (i.e., f/4) with which he could reach down to magnitude 15.5 on medium-format T-Max 400 film. He also occasionally used a 25-cm aperture f/2.8 Baker–Schmidt camera with a limiting photographic magnitude of 17. Takamizawa divided up the sky into 720 patrol regions with the 400-mm lenses, with each field covering roughly 8 degrees. He detailed his observing statistics at the 1999 International Workshop on Cometary Astronomy (IWCA) in August 1999. During the years 1994–1999 he observed on 367 nights (roughly one night in five) and checked 16,530 patrol photographs, roughly 45 per clear night. The checking of these had led to the discovery of 3 novae (1 recurrent), 1 supernova, 2 comets, and a colossal 502 new variable stars!

Figure 2.17. Japan's Kesao Takamizawa of Saku-machi, Nagano, Japan is shown here with a pair of 25 × 150 mm Fujinon comet-sweeping binoculars, although his main patrol instrument has been a system similar to that of his countryman Katsumi Haseda: that is, twin 400-mm-focal-length f/4 patrol lenses and 400 ISO film. In total he has discovered two novae, an outburst of the recurrent nova CI Aql, one supernova, two comets (one periodic: 98P/Takamizawa) and a colossal 502 new variable stars! The nova discoveries were V1425 Aql and V2487 Oph. Image by kind permission of Kesao Takamizawa, communicated to the author by Seiichiro Kiyota.

Blinking the Images

Perhaps the biggest advantage of using DSLRs for nova patrolling is the ease with which the master and patrol images can be 'blinked', or rapidly alternated on a PC screen, instantly revealing all the variable stars, asteroids, and the rare novae that are above the magnitude threshold. If the camera is piggybacked on a permanently mounted high-quality telescope mount, with accurate Go To slewing, the patrol field can be framed to a precision of a few arcminutes, making the precise alignment a very slick operation. In addition, unlike supernova patrolling, you do not need to align hundreds of images after a night's work. You may just have half-a-dozen images to check, making the whole checking process a lot more tolerable.

A number of astronomy packages will allow easy alignment of astronomical images (such as monochrome FITS format files), but the large color images captured by DSLRs cannot always be loaded into these applications. If a large image you have taken fails to load into an astro-package, there are a number of

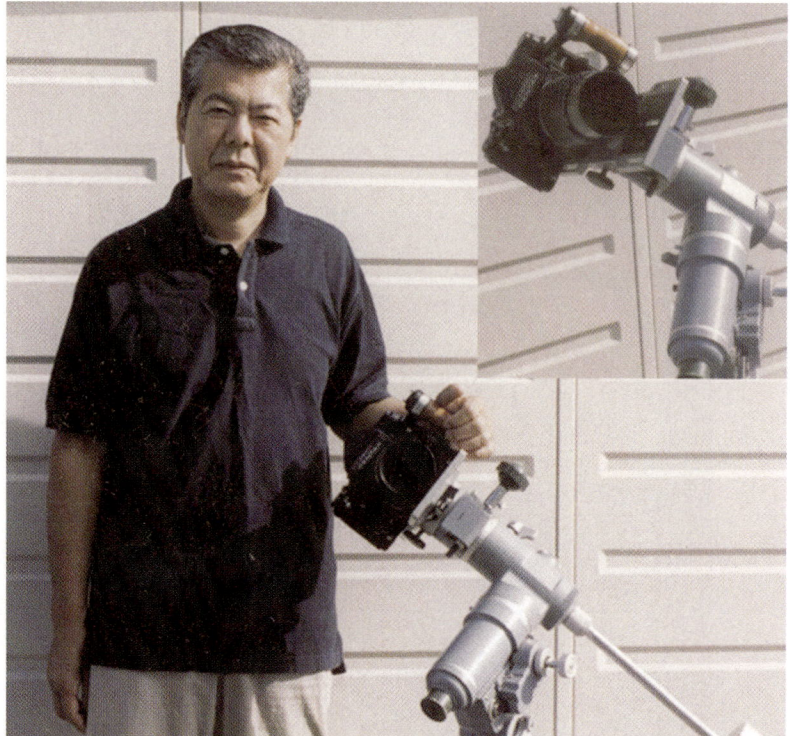

Figure 2.18. The prolific Japanese discoverer Hideo Nishimura of Kakegawa, Shizuoka, Japan, who discovered eight novae (including four co-discoveries shared with Liller, Nakamura, Sakurai and Pojmanski) of 8th–10th magnitude between 2003 and 2007. He uses a portable equatorial mount and a medium format Pentax 6 × 7 camera using T-Max 400 film and a 200-mm-focal-length f/4.0 lens, covering more than 15 × 15 degrees per frame. Nishimura's novae were discovered in Sagittarius (2), Cygnus (2), Scorpius (2), Ophiuchus, and Scutum. Image by kind permission of Hideo Nishimura, communicated to the author by Seiichiro Kiyota.

steps you can take. Firstly, you can try resaving the file in a different format. BMP, TIF, and JPEG images should all be compatible with most image-processing applications. If that fails, you can try reducing the image size; some older programs simply do not like multi mega-pixel images! Resizing the master and patrol images to 50 percent or smaller is easy in Photoshop or Paintshop Pro. If you still get no joy, converting the image to a grayscale (i.e., black and white) sometimes works.

An associated problem that sometimes complicates the blinking process is if two images are of slightly different size. This can happen if the image has had to be rotated by 1 degree or 2 degrees because your camera was tilted on its mount. When a rotation is applied and the image cropped, it may end up with a slightly different number of pixels. Careful cropping of the image, or increasing the background canvas size to a standard value, can sometimes solve this problem. This author has successfully used the 'align and blink' software in Richard Berry's AIP4Win software, which has a useful two-star alignment routine that successfully eliminates rotational differences. Software Bisque's CCDSoft is fairly slick at aligning and blinking images that have no rotational misalignment. Of course,

Figure 2.19. The world's greatest nova discoverer, Bill Liller (left), is pictured here talking to Dan Green of the Central Bureau for Astronomical Telegrams (CBAT), the astronomical clearing house for discoveries. This picture was taken by the author at the International Workshop for Cometary Astronomy II in Cambridge, England, on August 15, 1999.

downloading freeware to do the job can save you an expensive disappointment, as you might find the pricey package you purchased will not align and blink your images. Fortunately, most commercial packages now allow you to download a trial version, which expires after a month; this gives you time to test the software prior to purchase. The IRIS software package is, perhaps, the most popular astronomy freeware and can be downloaded from www.astrosurf.com/buil/us/iris/iris.htm.

IRIS is a partly command-line-driven system that some amateurs, used to a more graphics-oriented menu system, may not like. But it is powerful and was designed by amateurs for amateurs. For example, to quickly register two images simply by a horizontal and vertical alignment using one star in the field, you use the command-line QR [NAME1] [NAME2], where NAME1 and NAME2 are the image file names. Another basic freeware blink package is Andrew Sprott's Blink Comparator, which can be downloaded from the Trefach Astronomy Center's site at http://www.crysania.co.uk/trefach/trefach_astronomy.pl

Avoiding False Alarms

Unfortunately, there are plenty of objects in the night sky that can masquerade as novae, especially in the Milky Way. If you make a habit of announcing discoveries that turn out to be false, you will not be very popular. The same applies to supernova discovery claims. The galaxy is full of variable stars that brighten and fade slowly or explosively. A tiny proportion of them, usually the most violent, are covered in this book, but there are 40,000 in total and 10,000 more that are suspected of variation. Not surprisingly. the constellations that contain the most variable stars are also the ones that are most likely to produce the most novae, because their star density is very high. Fortunately, the vast majority of variables likely to catch the attention of nova patrollers, namely, those that peak above magnitude 11, are well documented.

Any high-quality planetarium package, such as Project Pluto's Guide 8.0 or Software Bisque's The Sky, will reveal variable stars when that object category is enabled and set to the same magnitude limit as the nonvariable stars. A conventional paper star atlas is invaluable, too. Wil Tirion's Uranometria 2000 is, perhaps, the most practical, as it reaches down to magnitude 9.5 and the charts each cover roughly 13 × 13 degrees, a very useful scale and equivalent to the vertical height of a 100-mm lens frame with a 36 × 24 mm image sensor. Of course, no paper chart can show you moving asteroids or comets, but any planetarium software will show the brightest asteroids. In addition, the webpages of the minor planets (at http://scully.harvard.edu/~cgi/CheckSN) will reveal any asteroids, too. Although faint asteroids in high-inclination orbits can appear at any declination, and lead to many false alarms for supernova patrollers, the nova patroller is somewhat more fortunate. In general the regions within 10 degrees of where the ecliptic plane crosses the Milky Way (i.e., in Sagittarius and Taurus/ Gemini) are the only areas where extreme caution is necessary.

The bottom line, where avoiding false alarms is concerned, is consulting with wise watchers of high repute before making any discovery claim. Two heads are better than one, and a few experienced heads are even better. If you cannot identify your nova suspect as a variable star, asteroid, comet, or already discovered nova, someone else probably can. Genuine nova discoveries are rare, and the Japanese and Bill Liller usually get them first, so extreme caution is necessary before contacting the CBAT (Table 2.7).

Patrolling for Novae Outside the Milky Way

Many amateurs, including this author, have discovered novae in external galaxies by using techniques and equipment more relevant to supernova patrolling. Novae are, of course, much fainter in absolute terms than supernovae, so the only sensible external galaxies worth patrolling, with modest amateur telescopes or lenses equipped with CCD cameras are M31 and M33 in the northern hemisphere constellations of Andromeda and Triangulum and the LMC and SMC in Dorado and Tucana in the southern hemisphere. The NGC designations for M31 and M33 are NGC 224 and NGC 598. The SMC is also cataloged as NGC 292. As far as this author is aware no amateurs are patrolling for novae in globular clusters, although plenty of deep-sky observers image these densely packed balls of stars every moonless night. A nova might have been discovered in one of the Milky Way's globulars. On May 21, 1860, Luther and Auwers of Berlin Observatoty discovered a 7th-magnitude nova, T Sco, in the well-known Messier globular M60 (NGC 6093). It had faded to 10th magnitude 3 weeks later. In a recent X-ray survey of M60 by the orbiting Chandra Observatory, five sources were identified as probable CVs, the brightest of which may be the X-ray counterpart of the 1860 nova T Sco.

Table 2.7. The world's most successful galactic (Milky Way) nova patrollers to 2007, not including extragalactic nova discoveries

Patroller	Discovery Years	Discoveries	Equipment	Location	Status
Liller	1983–2007	38	85-mm f.l. lens + film	Chile	Am
Honda	1970–1987	14	150-mm f.l. lens + Tri-X	Japan	Am
Haro	1949–1957	12	Pro. Astrograph	Mexico	Pro
Fleming	1897–1910	11	Pro. Astrograph	Harvard, U. S.	Pro
Woods	1897–1926	11	Pro. Astrograph	Harvard, U. S.	Pro
Camilleri	1991–1993	9	135-mm and 85-mm lenses	Australia	Am
Mayall	1936–1947	9	Pro. Astrograph	Harvard, U. S.	Pro
Nishimura	2003–2007	8	200-mm f.l. lens + film	Japan	Am
Plaut	1935–1959	8	Pro. Astrograph	Netherlands	Pro
Wolf	1904–1927	8	Pro. Astrograph	Germany	Pro
Cannon	1899–1926	7	Pro. Astrograph	Harvard, U. S.	Pro
Sakurai	1994–2007	7	180-mm f.l. lens + DSLR	Japan	Am
Hoffmeister	1928–1963	6	Pro. Astrograph	Germany	Pro
Kuwano	1971–1979	6	100-mm f.l. lens + film	Japan	Am
Pojmanski[1]	2004–2006	6	ASAS 200-mm + CCD	Chile/Poland	Pro
Nakamura	2001–2007	6	135-mm f.l. lens + CCD	Japan	Am
Alcock[2]	1967–1991	5 (+1)	Binocs (mainly 15 × 80)	England	Am
Haseda	2000–2005	5	135/400-mm lens + film	Japan	Am
Leavitt	1900–1906	5	Pro. Astrograph	Harvard, U. S.	Pro
Tago	1994–2007	5	105-mm f.l. lens + film	Japan	Am
Wakuda	1983–1988	5	400-mm f.l. lens + film	Japan	Am
Zwicky	1942–1949	5	Pro. Astrograph	Palomar, U. S.	Pro

[1]Pojmanski is Polish, but checks images from the automated ASAS (All Sky Automated Survey) patrol lenses based at Las Campanas Observatory, Chile.
[2]Alcock discovered five novae visually, but made an independent recovery of the recurrent nova RS Oph. The totals for Fleming, Leavitt, and Woods include outbursts of the recurrent novae RS Oph, T Pyx, and V1017 Sgr, although recurrent novae were not really understood in their era.

Andromeda and the Pinwheel

Up to 20 novae, typically of 16th–18th magnitude, are discovered in the Andromeda galaxy, M31, every year. This is far more than we see in our own Milky Way, but then as we live inside the plane of our galaxy its dust and gas will obscure many novae from our view. M31 is quite a big target, spanning 3 × 1 degrees, so a mosaic of images may well be needed to cover it with a telescope capable of clearly resolving 18th-magnitude novae. However, most of the novae are nearer to the central bulge of the galaxy. Unfortunately, this, in itself, causes problems. That central bulge can easily saturate the CCD detector in long exposures and needs to be tamed by reducing the contrast and brightness when searching for novae. In addition, carrying out astrometry (positional measurement) in such an over-exposed area, but with relatively few astrometric stars available, can be tricky. An associated problem is rooting out a database of massive variable stars within M31 that can masquerade as novae and fool the beginner. Admittedly, few are bright enough to shine as brilliantly as a nova, but they do exist. Fortunately there exists a General Catalog of Variable Stars (GCVS) catalog of extragalactic variable stars that can be found online at http://heasarc.nasa.gov/W3Browse/all/gcvseg-vars.html, which lists a total of 10,979 variable stars in 35 external galaxies and stellar systems. In addition the recent paper entitled: "Variable stars towards the bulge of M31: the AGAPE catalogue" (arXiv:astro-ph/0405402 v1 20 May 2004) is worth acquiring. Some professional patrollers of M31 employ an H-alpha filter for isolating extragalactic novae and suppressing the central bulge. These cut out a large percentage of the incoming light, as they typically have bandwidths of only 6 nm or 7 nm. However, in long exposures with sensitive CCDs any novae tend to stand out. Despite the problems of the overexposed bulge and the astrometry, novae in M31 have been discovered by four British observers, namely Ron Arbour, Mark Armstrong, Tom Boles, and this author. In my case the discovery was sheer luck! I was imaging another nova at the time and spotted the new one!

Apart from novae, obviously the discovery of a supernova in M31 would be a major event, with a lot of associated kudos for the discoverer. For a few hours on November 24, 1987, rumours escalated (and approached near-hysterical proportions) of a supernova in M31, until Lancashire amateur Denis Buczynski obtained a photograph showing that there was no such object.

M33, sometimes called the Pinwheel Galaxy, is face-on to us, and, conveniently, roughly a degree in diameter, but it only produces about two or three novae per year. At first this might seem surprising, as both the M31 and M33 galaxies are nearby (2.2 million and 2.4 million light-years, respectively), and M33 is face-on to us. However, M33 is a much smaller galaxy with maybe only 30 billion stars compared to an estimated 300 billion in M31.

For amateurs with very dark skies, owning large telescopes with sensitive CCD cameras, patrolling the galaxy M81 for novae is a possibility, too. A dozen or so novae between 20th and 23rd magnitude are discovered in that large galaxy each year by professional astronomers. The entire galaxy can almost be squeezed into a 15-arcminute-wide CCD frame, which is convenient, but only if you can go really deep. I do not know of any amateur discoveries of novae in M81. Determining the rate at which novae occur in external galaxies can help to refine the relationship between galaxy types and stellar populations. It is interesting to note that at the time of writing Liverpool John Moores University was carrying out a survey called Liverpool Extragalactic Nova Survey (LENS), in which it was regularly patrolling three relatively nearby galaxies, namely M81, M64, and NGC 2403 using the 2-meter Liverpool (LT) and Faulkes Telescope North (FTN) instruments. The aim is to discover 50–100 extragalactic novae to better understand extragalactic nova behavior and population types and understand how these novae can be used accurately as distance indicators. Luminous hot and young stars, frequently found in a galaxy's spiral arms, are classed as Population I, whereas Population II stars are older, less luminous, and found in the central regions of galaxies and in globular clusters. Population I stars contain more heavy elements, and these heavy elements are thought to be formed in supernovae from a previous generation of stars.

The LMC and SMC

For southern hemisphere amateur astronomers the LMC and SMC, being far closer to us than even M31 and M33, enable an extragalactic nova patrol to be carried out using equipment as modest as a telephoto lens. The LMC is only 170,000 light-years away, and the SMC roughly 210,000. Thus, novae in these companion dwarf galaxies to our own Milky Way typically peak at magnitude 11 or so, ideal for coverage with a 200- or 300-mm lens and a DSLR, with exposures of 1 or 2 minutes. Such lenses, with a full-frame CCD sensor, will have fields of view between 5 degrees and 10 degrees, which will be well matched to the LMC and SMC dimensions. Between 1988 and 1990, Australian amateur astronomer Gordon Garradd discovered four novae in the LMC using a 300-mm-focal-length f/4.5 lens and gas-hypersensitized Kodak 2415 film. A modern DSLR system with the same lens would be just as sensitive, and, of course, the images could be checked as soon as they were exposed. On average the LMC produces one or two novae each year, which is roughly in keeping with its likely population of 10 billion stars. With an estimated population of only a few billion stars the SMC is a rather poor source of novae, but patrolling it alongside the LMC would be a logical approach.

Paul W. Hodge of the University of Washington has compiled a number of books about nearby galaxies that may be invaluable to potential extragalactic nova hunters. These include atlases of M31 and Local Group galaxies. (See the Resources section in this book for more details.)

Obtaining Spectra of Novae

The telescopes and CCDs in amateur hands are ideally suited to obtaining useful spectra of bright novae, but few amateurs seem to be engaged in this activity. This, of course, makes it all the more valuable work. When a bright naked-eye nova appears, the professional observatories can sometimes find it is too bright for their equipment, leaving amateur spectroscopists in the ideal situation. When the light from a nova passes through the outbursting star's gas on its way to Earth, photons are going to be absorbed by electrons orbiting the atomic nuclei in the gas and the electrons will then jump up a discrete orbit level as they absorb the photons. This absorption leads to dark lines appearing at discrete wavelengths in the spectrum. The opposite effect occurs when the gas is being excited by some energy input. In this case the electrons may jump down a discrete orbit level as they emit photons. This emission leads to bright lines appearing in the spectrum. The well-known Balmer series of hydrogen atom orbit transitions give rise to lines in the visible part of the spectrum and correspond to electron transitions between the second orbit level and higher orbit levels. It was only with the development of quantum physics that the discrete allowable orbits were fully understood.

Figure 2.20 is a block diagram of the basic components of a single-prism spectrograph. Although this may look daunting, the components are not hard to acquire, especially if you contact other amateurs who are into spectroscopy and can give you a headstart. A few websites for learning more about amateur spectro-scopy are listed in the Resources section of this book. The goal of the activity is to collect as much light as possible from the star but not from anything else, and then split the star's light up into a spectrum and focus it. You need to channel parallel light from the star through a prism and then use a lens to focus the red end of the spectrum at one end of the CCD detector chip and the blue end at the other. This is the simplest, most efficient, and practical way to capture the spectrum.

In practice, the prism can be replaced with a diffraction grating, which disperses the light in a slightly different way. With a diffraction grating, dispersions are conveniently greater than with a single prism (older spectroscopes often used several prisms in sequence), but they produce two sets of spectra, each with several

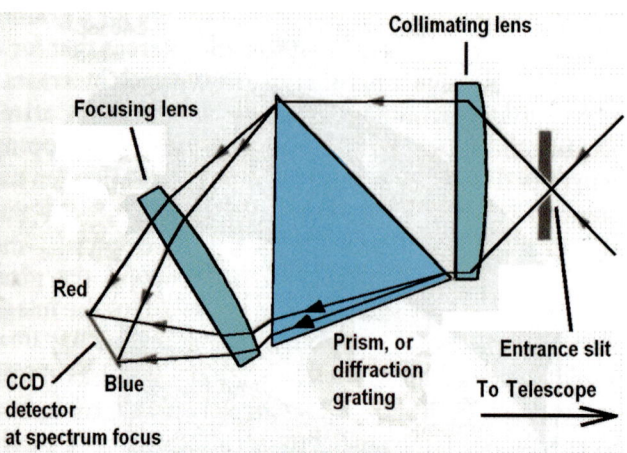

Figure 2.20. The basic components of a spectrograph. The purpose of the design is to collect as much light as possible from the star, but not from anything else (hence the slit), and then make the beam parallel as it enters the prism. This splits the star's light up into a spectrum, and the focusing lens places focused red light at one end of the CCD and focused blue light at the other. In practice the prism can be replaced with a diffraction grating, which disperses the light in a slightly different way.

'orders' of spectra; the majority of the light goes into the white-light 'zero order' spectrum. Thus, the spectra are not as bright. However, if the grating is of the 'blazed' type (more often found in reflection gratings), the individual grating line surfaces are angled to direct the majority of the light into the spectrum. Obviously, to take advantage of this, the grating has to be angled accurately to direct the bright spectrum at the detector. Every spectroscope starts with a slit. In a normal telescope this is where the eyepiece would focus or the CCD would be placed, that is, the focal plane, where the image of the star field exists. The purpose of the slit is to reduce background noise from the rest of the sky and reduce any overlap from adjacent wavelengths. The narrower the slit, the better the spectrum is resolved, but, if the slit is narrower than the focal plane star diameter, light will be lost.

The collimator is simply a lens designed to ensure that parallel light enters the prism. Once the parallel light has been split into a spectrum by the prism, the spectrograph's own lens, the imaging lens, easily focuses the red light to one end of the CCD and the blue to the other; so the spectrum is nicely spread out along the chip.

The next issues to be addressed are how well can the spectrum be resolved, what focal length should the spectrograph telescope lens be, and how much of the spectrum will fit onto the length of the CCD? With the typical prisms or gratings available to amateurs, the middle of the visual spectrum can be resolved as finely as 1 Å. But these same prisms or gratings typically disperse the spectrum such that 1 Å of the spectrum subtends an angle of, say, only 2 arcseconds. Thus, the spectrograph's imaging lens will need a focal length of a meter to capture 1 Å of resolution per 10-micron CCD pixel. At this scale, however, a 500-pixel-long CCD array will only capture 500 Å of visual spectrum, compared to the whole visual spectrum of 4000–7000 Å, that is, 3000 Å. A spectroscope will have a spectral

Figure 2.21. The spectrograph of Robin Leadbeater, fixed to the back of his 280-mm-aperture Celestron 11. An ATIK ATK16-IC camera is used to record the spectra of a nova or variable star while a modified webcam is used as an autoguider. The spectrograph is the LHIRESIII unit developed by the French AUDE Astronomy Group and sold in kit form. (www.astrosurf.com/thizy/lhires3/index-en.html.) The unit is fitted with two diffraction gratings, one of 2400 lines per millimeter resolution and a lower resolution one of 150 lines per millimeter. The lower resolution grating is used for fainter objects. Image by kind permission of Robin Leadbeater.

resolution set by the diffraction grating's performance, but this can be compromised if the slit is widened (to reduce exposure times) and by instrument deficiencies. However, to actually capture the resolution on the CCD, the dispersion and the focal length of the imaging lens/mirror must deliver a small enough 'angstroms per pixel' scale. It is all a bit of a balancing act!

With bright naked-eye novae even a small telescope can be used to produce a spectrum with an exposure time of under a second. For the DIY spectroscope builder, optimum grating/prism assemblies are rarely available – likewise, for the collimating and imaging lenses. It is usually a case of buying cheap components and bolting them together to see what happens, something of a hobby in itself. Amateur spectrographs are rarely designed precisely. Fortunately, diffraction gratings of 600 linesper millimeter can be purchased for as little as $25 and adjustable slits can be made from two razor blades. In addition, secondhand camera lenses can be called into service for the collimating and imaging lenses, leaving the CCD as the most expensive component. But there are other technical considerations, too. For example, how do you actually keep the telescope guided so that the star being analyzed is kept in the slit? One way of doing this is to focus a guiding eyepiece or telescope on the outer surface of the slit; this surface, if highly polished, will easily show the outer overspill of the stars disk. It is actually advantageous to let the star's right ascension drift trail back and forth along the slit length, as this produces the height of the spectrum. With perfect tracking the spectrum would be an almost infinitesimally thin line and very hard to analyze. For those with deep pockets and a fascination with spectroscopy a few thousand dollars will secure you a professional spectrograph from SBIG.

A nova's spectrum changes over time as the star peaks in brightness and fades. Some recent novae have exhibited some fascinating standstills and fades in their light-curve that can only be explained when coupled with an analysis of the spectrum. Figure 2.21 shows Robin Leadbeater's spectrograph and Figure 2.22 shows his inexpensive Paton Hawksley Star Analyser attached to an ATiK CCD camera. One of Robin's spectra is featured in Figure 2.23.

There are commercial and freeware packages that can take the image of a spectrum and convert it into a graph for use in a spreadsheet. The package Visual Spec is very popular and can be found at http://astrosurf.com/vdesnoux/. Richard Berry's AIP4Win has a number of spectroscopy features, too: http://www.willbell. com/AIP/Index.htm.

Figure 2.22. The Paton Hawksley Star Analyzer, seen here attached to Robin Leadbeater's ATIK CCD camera, is an inexpensive ($150) way of obtaining spectra using a webcam on bright stars or a cooled CCD camera on novae or even bright supernovae. It features a high-efficiency 100 lines per millimeter blazed diffraction grating and screws into any 31.7-mm filter thread. With an 80-mm telescope and a simple, unmodified webcam, stars down to magnitude 4 can be analyzed. www.patonhawksley.co.uk/staranalyser.html. Image by kind permission of Robin Leadbeater.

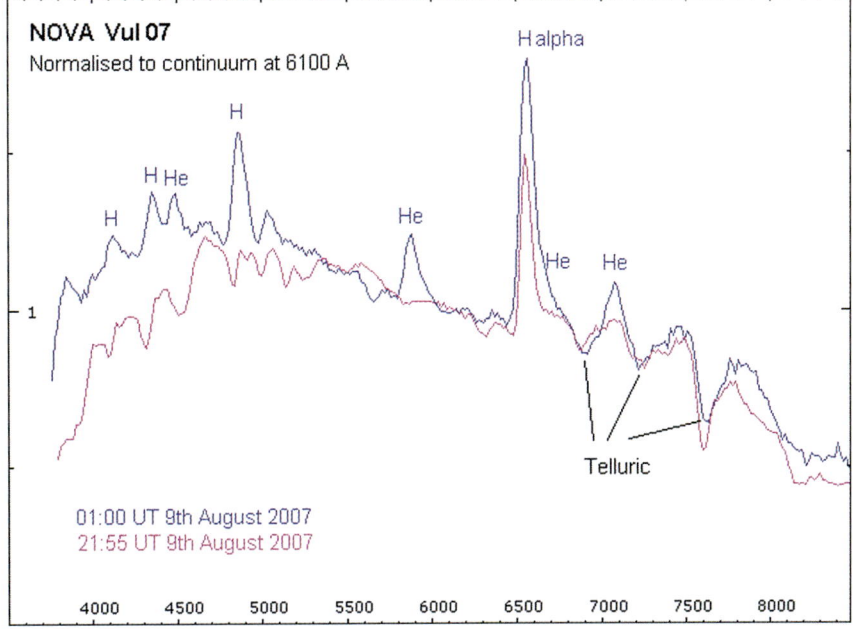

Figure 2.23. Two spectra of Nova Vulpeculae 2007, obtained 21 hours apart, by Robin Leadbeater on August 9, 2007, using a 200-mm-aperture Vixen VC200L and Star Analyser diffraction grating. Image by kind permission of Robin Leadbeater.

Checking for Recurrent Nova Outbursts

From a patroller's point of view, checking for outbursts of recurrent novae is no different from checking for outbursts of dwarf novae; indeed, we have already included many 'NR' category objects in Table 1.2 in Chapter 1. Quite a few objects that have only been seen in outburst once, or twice, have not had their classification pinned down. Spotting an outburst of one of these objects carries a huge amount of prestige and the knowledge that professional astronomers will be trying to swing large telescopes, and even spacecraft observatories, onto the objects in question. Once again finder and magnitude comparison charts can be acquired or generated from the AAVSO websites at www.aavso.org/observing/charts/ and http://www.aavso.org/observing/charts/vsp/.

Many of the objects in this category are probably unique in their behavior. However, there is one minor problem here. Of the nine cast-iron, proven, recurrent novae, all but one (T Corona Borealis) lie in the southern celestial hemisphere, although, admittedly, CI Aquilae and RS Ophiuchi are only just south of the celestial equator. So, patrolling the definite recurrent novae is largely a hobby for southern hemisphere dwellers.

However, there are plenty of unusual objects (like V404 Cygni) to keep northern hemisphere observers busy. It might be thought that simply checking for outbursts of these rare outbursting novae was a fairly dull pursuit! After all, the chance of a recurrent nova being in outburst when you examine the field is pretty slim. Although this is very true, and others with clearer skies and more spare time will

Table 2.8. Proven (light gray) and unusual or suspected (dark gray) recurrent novae

Star	Coordinates	Dist. (ly)	Outburst Years	Mag. Range
CI Aql	18 h 52m 03.59s –01° 28′ 39.3″	5000	1917, 2000	8.8–15.6
SV Ari	03 h 25m 03.34s +19° 49′ 52.9″	N/A	1905 and maybe 1943?	12.0–22.0
V394 CrA	18 h 00m 25.97s –39° 00′ 35.1″	13,000	1949, 1987	7.2–18.8
T CrB	15 h 59m 30.16s +25° 55′ 12.6″	2500	1866, 1946	2.0–11.3
V404 Cyg	20 h 24m 03.78s +33° 52′ 03.2″	5000	1938, 1989 X-ray Black hole	11.0–20.5
AY Lac	22 h 22m 22.10s +50° 23′ 40.0″	N/A	1928, 1962	14.5–<20
HR Lyr	18 h 53m 25.05s +29° 13′ 37.8″	N/A	1919	6.5–16.5
IM Nor	15 h 39m 26.47s –52° 19′ 18.0″	20,000	1920, 2002	7.8–~22
RS Oph	17 h 50m 13.16s –06° 42′ 28.5″	5000	1898, 1933, 58, 67, 85, 2006	4.3–12.5
V2487 Oph	17 h 31m 59.8s –19° 13′ 56.0″	N/A	1998 (plus 1900? See text)	9.5–17.7
UW Per	02 h 12m 29.54s +57° 05′ 18.5″	N/A	1912 (+1915, 17, 19, 47?)	13.5–21?
T Pyx	09 h 04m 41.51s –32° 22′ 47.6″	6000	1890, 1902, 20, 44, 66	6.3–15.3
U Sco	16 h 22m 30.81s –17° 52′ 44.1″	20,000	1863, 1906, 17, 36, 45, 69, 79, 87, 99	8.8–19.5
V745 Sco	17 h 55m 22.25s –33° 14′ 59.5″	16,000	1937, 1989	11.2–~21
EU Sct	18 h 56m 13.17s –04° 12′ 33.4″	N/A	1949	8.4–~18
FS Sct	18 h 58m 16.93s –05° 24′ 05.2″	N/A	1960	10.1–19.3
V3645 Sgr	18 h 35m 49.21s –18° 41′ 45.1″	N/A	1970	12.6–~18
V3890 Sgr	18 h 30m 43.32s –24° 01′ 08.6″	13,000	1962, 1990	8.4–17.2

probably beat you to the discovery, when included in a nightly check of dozens of other CVs the statistics are less bleak. In addition, many of these objects show variability at the low end of their magnitude range, and this allows useful magnitude estimates to be made visually down to 15th and 16th magnitude (with a large enough instrument) and considerably lower with CCD detectors (Table 2.8).

The suspected recurrent novae in this table fall short of being classed as proven recurrents for various reasons. Some, like SV Ari and UW Per, may have outburst more than once, but the evidence is flimsy. Others, like AY Lac, may be recurrent novae, but a further outburst in the modern era would be needed to spectroscopically classify the outburst. V404 Cyg was originally classed as a nova in 1938, but when a further outburst occurred in 1989 it coincided with an X-ray outburst, and a black hole was seen to be the culprit. So V404 Cyg was classed as a nova, and then it recurred, but this system is highly unusual. Nevertheless a further outburst is well worth keeping a lookout for. Many of the objects in Table 2.8, in particular V2487 Oph, V3890 Sgr, V745 Sco, T CrB, U Sco, V394 CrA, and RS Oph are thought to have white dwarf primaries of around 1.35 solar masses; that is, they are perilously close to the 1.4-solar-mass Chandrasekhar limit, so may one day become Type Ia supernovae. Even if they do not, the extra mass and gravitation of these big white dwarfs provides a hair-trigger for nova eruptions after relatively modest accretion onto the primary in timescales of between 10 years and 80 years.

Recurrent Novae Worth Monitoring

Images of some of these objects appear in this chapter, but I have also provided a few thumbnail 5 arcminute field images in the Resources section of this book.

RS Oph and T Corona Borealis

RS Ophiuchi and T Coronae Borealis are the best known and brightest examples of proven recurrent novae in the night sky. Lying just less than 7 degrees south of the celestial equator, RS Ophiuchi has outburst on six separate occasions, in 1898, 1933, 1958, 1967, 1985, and 2006. With the star outbursting to magnitude 4.3 at its peak, being well monitored, and having produced an outburst as recently as 2006 (see Figure 2.24), another imminent outburst is unlikely until 2015 at the earliest (the 1958–1967 outburst interval was 9 years). However, with a minimum magnitude of 12.5 the star is easy to scrutinize even at its low point, and with binoculars being a perfectly reasonable patrol instrument for this recurrent nova, it is easy to check, and well worth monitoring, the field. George Alcock, famous for his five novae and five comet discoveries, often considered his independent detection of the 1985 RS Ophiuchi outburst as qualifying him as a six-nova man (it would have been his fifth nova discovery, making his 1991 Nova in Herculis his sixth). He spotted RS Ophiuchi in outburst at magnitude 6 on January 30, 1985, using hand-held 15 × 80 binoculars looking through a double-glazed indoor window! The Ontario observer W. Morrison had already reported it, but Alcock had no knowledge of that. Guy Hurst confirmed the outburst, also using 15 × 80 binoculars. Alcock missed bagging the 1967 flare-up of RS Ophiuchi due to driving home from Stowmarket to Peterborough in England. He had been visiting J. P. Manning-Prentice, the discoverer of the naked-eye nova

Figure 2.24. The recurrent nova RS Ophiuchi in outburst, imaged by Giovanni Sostero and Ernesto Guido on February 13, 2006. They used a 0.25-m f/3.4 Takahashi astrograph at the New Mexico Skies observatory, remotely, over the Internet. North is up and the field is roughly half-a-degree high.

Nova Herculis 1934. The 2006 outburst of RS Ophiuchi was first detected on February 12 by the Japanese observers Hiroaki Narumi (located at Ehime, Kita-gun) and Kiyotaka Kanai (located at Gunma, Isezaki-shi), who found it at magnitudes 4.5 and 4.6, respectively. The recurrent nova RS Oph has a much longer orbital period than a standard dwarf nova. Astronomers think the period is just under 456 days. The components are believed to be a massive white dwarf with a mass of 1.4 solar masses, just a shade below the Type 1a supernova Chandrasekhar limit. The secondary star is believed to be a large red giant of (spectral class M2-3 III).

T Corona Borealis is the brightest recurrent nova known, but only two outbursts have been spotted, in 1866 and 1946. Being such a bright (2nd magnitude) object in outburst, it is hard to believe that any outbursts could have been missed in the intervening years, unless they occurred when the star was hidden in the solar glare. Surely, when it was high in a dark sky, as in the northern hemisphere spring, an outburst could not have been missed! Well, maybe not, but as well as the solar glare issue there is also the speed of T CrB's decline. In a week it drops 3 magnitudes below its peak, and its peak was only 3rd magnitude at the 1946 outburst. So, with a week of cloud, it can quickly fade well below naked-eye visibility.

You will not even need binoculars to check if an outburst of T CrB is occurring at its peak; the naked eye is good enough! However, you may wish to check the binocular or telescope field for any odd behavior, and charts, as always, can be acquired from the AAVSO website. Like RS Oph, T, CrB has a long orbital period as CVs go. In this case a period of just less than 228 days fits the observations, and yet again, a massive white dwarf (around 1.35 solar masses) is the primary component. The secondary is thought to be a bloated red giant star with a mass of around 0.7 solar masses. The system's orbital inclination is probably close to 70°. The dedicated and legendary Ohio amateur astronomer Leslie Peltier rarely missed a clear night, but famously slept through the February 9, 1946, outburst of T CrB, despite meticulously looking specifically at the constellation for 25 years. He felt like he might have had a cold coming on and, as he described in his autobiography *Starlight Nights*:

> Self-pity comes easy at 2:30 on a cold February morning so I went back to my warm bed with the comforting thought that I owed it to my family, at least, to take care of my health. And thus I missed the night of nights in the life of T Coronae. It was the night spectroscopists long for. It is in those earliest hours of awakening that the newborn star – with all the exuberance of youth, divulges its most intimate secrets.

Peltier made it clear that his relationship with the star would never be the same after it flared while he was not observing it:

> I am still watching it but now it is with wary eye. There is no warmth between us any more.

One specific object that springs to mind, and one with which this author was closely involved, was the object known as V404 Cygni. So I will now indulge, Peltier-like, in a very personal recollection of the events of May 1989.

V404 Cyg

In 1989, I found myself in the right place at the right time when an old nova of 1938 went into outburst. The nova V404 Cyg was originally discovered by Wachmann at Hamburger Sternwarte Observatory on October 14, 1938, at magnitude 12.5. It probably peaked in magnitude at around 11.5 somewhere between September 28, 1938, and October 14. The star subsequently faded to at least magnitude 20.5 and was not seen in outburst again, until May 22, 1989. On this day I was just starting a week's holiday; a U.K. 'Bank holiday' week, in fact. I had e-mailed Guy Hurst of *The Astronomer* magazine, informing him that I was on holiday and if anything unusual was discovered, could he please let me know? Maybe I was clairvoyant on that occasion?! On May 22 Sagamihara reported the discovery of a bright X-ray outburst by the Ginga satellite. IAU Circular 4782 reported the likely position as 20 h 23.3m +33 50′ (1950), but with an uncertainty of around 10′. *The Astronomer's* E-Circular 303, issued by Guy Hurst on May 25, advised patrollers to check for an optical counterpart. In a telephone call, Guy advised me that Brian Marsden of CBAT thought it could be linked to the 1938 nova. Just before midnight on May 26/27 I had a crystal clear sky and took a 15-minute exposure of the V404 Cygni field. A few months earlier I had purchased a copy of Hans Vehrenberg's massive paper "Atlas Stellarum," primarily because it had an image scale of 2′ per mm, identical to the focal plane scale of my 0.36-m f/5 Newtonian. Thus, by overlaying my negatives onto Stellarum, checking for new objects would be simple, and so it proved. The new 12.5-magnitude object was immediately obvious, and, using the Stellarum grid, I e-mailed a position to Guy, along with a crude diagram, showing stars as full-stops or asterisks! (E-mail was in its infancy in 1989.). Of course, almost 20 years later, things are different; with CCDs and Hubble/Hipparcos CD-Roms, checking out a magnitude-12.5 object would be routine, and I could have e-mailed my image to Guy. But in 1989 my powerful checking technique, clear skies, new fangled e-mail facility and contact with Guy, meant that for several days I was the only amateur photographing this object; (Figure 2.25) no one else had any charts for it! On May 27, after Guy had relayed my results to Brian Marsden,

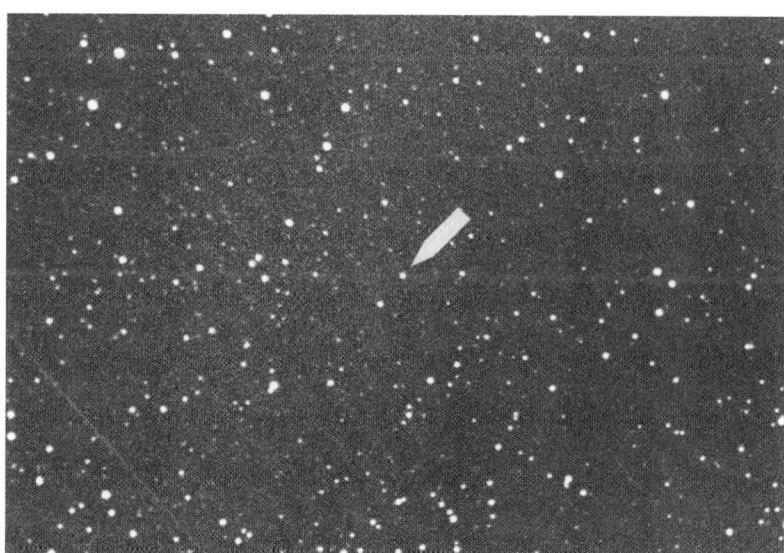

Figure 2.25. The recurrent object V404 Cygni, photographed by the author with a 0.36-m f/5 Newtonian and T-Max 400 film on May 30, 1989, from 23:06 to 23:21 UT. The magnitude of V404 Cyg on this image is approximately 12.6. The field is 20′ × 28′ with north at the top. V404 Cyg is an unusual X-ray nova whose primary appears to be a massive black hole.

an IAU Telex was issued confirming that the 1938 nova V404 Cygni was identical to the Ginga X-ray object. I cannot pretend that I was not somewhat smug that my modest observatory and the multi-million dollar very large array (VLA) appeared side by side as the principal facilities studying this object!

Subsequent studies have shown that V404 Cygni is one of our most likely black hole candidates, with a low-mass red giant of just under 1 solar mass orbiting the 10-solar-mass black hole. The orbital period of this highly unusual recurrent 'nova' is thought to be 6.5 days. Large amounts of material spiraling into the black hole's intense gravitational field are heated to 10 million degrees Kelvin, at which temperature X-rays are emitted. Clearly, V404 Cygni is far from being a normal nova, recurrent nova, or even a CV. But it is a unique system that amateurs can clearly monitor for unusual outburst activity.

HR Lyr

This star was first seen in outburst by J. C. Mackie, on Harvard Observatory plates taken in December 1919. It peaked at a very respectable magnitude 6.5 on December 6 and had faded to magnitude 9.5, 10 weeks later. The star was designated as Nova Lyrae 1919. Its minimum state appears to be at about 15.8, although it was reported as being below 16.5 on plates taken a few days prior to the 'nova' eruption. It has not experienced another outburst to magnitude 6.5 in the intervening decades, but there is considerable fluctuation in its 'quiescent' low-state brightness, on timescales as small as hours and as long as years. This may mean that it is actually a recurrent nova and could outburst again. Undoubtedly the Japanese nova patrollers would bag it immediately if it reached magnitude 6.5, but lesser outbursts are worth monitoring to determine its true nature.

T Pyxis

T Pyxis is definitely a recurrent nova for the southern hemisphere patroller, lying, as it does, at a declination of −32°. To date there have been five outbursts between 1890 and 1966, but, mysteriously, nothing since then. This recurrent nova is roughly 20 years overdue for another outburst!

Could an outburst have been missed? Maybe. The outbursts only peak at magnitude 6.3, so T Pyxis at maximum does not catch the eye, but nova patrollers would still have been expected to pick it up. However, more likely than a nova outburst simply being overlooked is that it might have been lost in the solar glare, near conjunction. Either way, another outburst soon is in the cards, so if you can see the constellation of Pyxis, make sure you check the region out. The last outburst, in 1966, was picked up by the legendary variable star observer and comet discoverer Albert Jones of Nelson, New Zealand. Albert had been checking the T Pyxis field since 1954, but on December 7, 1966, he saw it for the first time, at magnitude 12.9. By December 9 it had risen to 9th magnitude, and on January 11, 1967, it peaked at magnitude 6.3. Thereafter it dropped by 3 magnitudes in a 100 days, a further 2 magnitudes in the next 20 days, and eventually settled down to 15th magnitude. Southern hemisphere amateurs followed the decline with great interest but, as of 2008, have been kept waiting for more than 40 years for a repeat performance.

V3890 Sgr

Close to the galactic center, the field of the southern hemisphere recurrent nova V3890 Sgr is well worth tracking down, especially if you live at the latitude of South Africa, Australia, or New Zealand. With two outbursts being observed in relatively recent times, only 28 years apart (1962 and 1990), and the possibility that an intermediate outburst might have been missed in this crowded part of the Milky Way (surely there must have been other outbursts in the photographic era prior to 1962?), this is a top-priority object. Its quiescent magnitude of around 17.2 also allows CCD monitoring at the low end of its range, in case any fluctuations occur at that minimum state. The original outburst was found by H. Dinerstein on plates taken by D. Hoffleit at the Maria Mitchell Observatory. The maximum magnitude observed was 8.4, with the peak occurring between May 10 and June 2 in 1962. The 1990 outburst was spotted, again by the veteran observer Albert Jones from New Zealand. Jones found it to be at magnitude 8.5 on 1990 April 27.72 U.T.

U Sco and V745 Sco

A couple more southern hemisphere nova fields worth a routine check are those of U and V745 Sco. At minimum these two objects would be a test even for modest-aperture CCD systems, as they sit at magnitudes 19.5 and 21, respectively. Even at maximum they are hardly spectacular (magnitudes 8.8 and 11.2). Nevertheless, proven recurrent novae are rare, so we cannot afford to be too fussy. U Sco is sometimes reported as having outbursted six times, with its initial outburst of 1863 and five outbursts later. However, its outburst has been recorded at least nine times, making it the most prolific known recurrent nova, even if it cannot compete in brightness terms with T Cor. B or RS Oph. As recently as February 2004 B. E. Schaefer of Louisiana State University announced that a previously unknown eruption of U Sco had been discovered on Harvard College Observatory archival photographs of March 6, 1917, when it appeared to have a blue magnitude of 9.1. Schaefer noted that U Sco had a fairly constant recurrence cycle of 8–12 years – with about 25 percent of the outbursts being missed due to proximity to the Sun (including potential missed outbursts in 1926 and 1957). He added, on IAUC 8279, that the next U Sco eruption should occur sometime during 2007–2011.

U Sco was initially discovered by Pogson in 1863, with the other outbursts taking place in 1906, 1936, 1979, 1987, and 1999. Thus, the average outburst interval, based on the last three outbursts, seems to be about 10 years, and the next outburst could be due any time soon. The 1987 outburst was captured by the late South African amateur astronomer M. Daniel Overbeek (1920–2001) on May 16 of that year. The 1999 outburst was caught by Patrick Schmeer on February 25, 1994 U.T., when it was seen at magnitude 9.5. Less than 4 hours earlier the skilled observer (and, more recently, supernova discoverer) Berto Monard had found that the star was invisible to a limit of magnitude 14.3. U Sco peaked quickly, that same day, at magnitude 8.0, but only a week later it had faded back to 11th magnitude. This emphasizes just how easily such outbursts can be missed, and why a worldwide network of observers is so vital, if only to get around the cloud that seems to accompany any time-critical astronomical events.

V745 Sco has only outburst twice, in 1937 and 1989, but the 52-year gap may have contained a few missed outbursts. With these distant and rather faint examples of recurrent novae this is quite possible, especially prior to the 1990s, when Go To instruments were not available to amateurs. The veteran nova discoverer Bill Liller, based at Vina del Mar, Chile, bagged the 1989 outburst on July 24 of that year, using an 85-mm-focal-length lens and Kodak 2415 film.

V394 CrA and IM Nor

V394 CrA was first seen in outburst by L. E. Erro at Tonantzintla Observatory on March 23, 1949, as a magnitude 7.5 star. On plates exposed on the night before no object brighter than magnitude 12.5 was visible. Because the nova was dead on the border between Crater and Scorpius it was called Nova CrA 1949 by some and Nova Sco 1949 by others! Thirty-eight years and just more than 4 months then elapsed until, on 1987 August 1.997 U.T. Bill Liller (who else?) from Vina del Mar, Chile, photographed it at magnitude 8.9 with an 85-mm f/1.4 lens. Another 38 years will be up in 2025, but that should not deter keen southern hemisphere patrollers from regularly checking the field.

Fifteen years later Liller swept up yet another recurrent nova of the far southern skies, the farthest southern example as far as we know, as IM Normae resides at 52 degrees below the celestial equator. IM Normae is another recurrent nova that has only been seen in outburst on two occasions. The original outburst was discovered by I. E. Woods at Harvard Observatory as a 9th-magnitude star on a photographic plate exposed on 1920 July 7. At minimum the star sinks to magnitude 22, so it was right on the limit of the most powerful telescopes of that photographic era. In 1972 Liller (working as a professional astronomer) and J. L. Elliott suggested (ApJ 175, L69, 1972) that IM Normae was an X-ray source, and Liller went on to detect its second outburst on January 10, 2002, at magnitude 8.3. So, with 82 years between outbursts, do not get too excited by the prospects this one offers, but, as always, it is definitely worth incorporating into a systematic patrol if you live in the southern hemisphere.

V2487 Oph and V3645 Sgr

On June 15, 1998, the prolific Japanese amateur discoverer Kesao Takamizawa reported the discovery of a magnitude-9.5 nova in Ophiuchus on patrol negatives taken with his twin 400-mm lenses. No star was present in 48 previous patrols he had taken of that region between February 17, 1994, and May 19, 1998. The nova would ultimately be designated as V2487 Ophiuchi. The star faded by 3 magnitudes in only 9 days (one of the fastest nova decline rates ever seen) and then stayed virtually constant (a plateau) for the next 3 weeks. This light-curve behavior is virtually identical to that of the recurrent nova U Sco (see earlier section), which has outbursted six times with a typical interval of 10 years.

At the time of writing it is 10 years since the first detected outburst of V2487 Oph, so keep an eye on this one. It normally sits just above magnitude 18. As we have seen, professional astronomers who have studied the known recurrent nova systems think that many are potential Type Ia supernovae because the white dwarf primary stars are very close to the critical 1.4-solar-mass Chandrasekhar limit.

V2487 Oph, V3890 Sgr, V745 Sco, T CrB, U Sco, V394 CrA, and RS Oph are all thought to be very close to the edge in this regard, with masses of perhaps 1.35 solar masses, and increasing as much as a tenth of a millionth of a solar mass per year. A quick calculation tells us that if this is true, they are less than half a million years from changing from recurrent novae to Type Ia supernovae. Understanding these systems in outburst is therefore crucial to nova and supernova theory. V2487 Oph was detected with the XMM-Newton telescope in 2000 and 2001, which showed that accretion was re-established less than 3 years after its outburst and hinted that it was behaving like a magnetic CV, which was rather puzzling for a nova. Clearly this is a mysterious object worthy of further study. As this book was going into production A. Pagnotta, et. al., Louisiana State University, reported their possible discovery of a previously unknown eruption of V2487 Oph in 1900 in the Harvard College Observatory archival photograph collection.'

V3645 Sgr was discovered in 1970, roughly 6 months after it peaked in brightness. V. Archipova and O. Dokuchaeva found it on an objective prism plate, taken by R. Bartaya and T. Vashakidse at Abastumani Observatory on July 29, 1970. The nova reached only 12.6 in outburst but faded very slowly (3 magnitudes in 10 months). It probably peaked at 12th magnitude during the winter months of 1969/1970. In a 1994 paper published in the RAS Monthly Notices (Vol 266) by Weight, Evans, Taylor, Wood (from Keele University), and Bode (Liverpool John Moore's University), V3645 Sgr was named as being a potential recurrent nova, as it only appeared to have a range of less than 6 magnitudes and was an unusually red object. This red color may be due to the presence of a giant secondary star. There is certainly much to learn about this object, not the least because its 1969/1970 outburst went virtually unknown and an outburst light-curve is not available. This is definitely an object worth careful monitoring.

CI Aql

Along with T Corona Borealis and RS Ophiuchi this is only the third proven recurrent nova (out of nine) that is easily observable from northern Europe or North America. With 83 years between known outbursts (1917 and 2000), it would seem unlikely that a third outburst was due any time soon; but if it is, and with Aquila being such a fertile producer of classical novae that it is scoured remorselessly by Japanese nova patrollers, a recurrence will probably not be missed. The only likely scenario preventing an outburst being spotted is when the region is too close to the solar glare, such as at the start of the year.

With a minimum magnitude of 15.6 this is another object that can be bagged at minimum by the sharpest visual patrollers using large amateur telescopes. The 2000 outburst was detected by Kesao Takamizawa, who we discussed earlier, and who this author had the pleasure of meeting at the 1999 International Workshop on Cometary Astronomy in Cambridge, England, just after the August 1999 Total Solar Eclipse. Takamizawa is a successful nova and comet discoverer, primarily using twin 400-mm-focal-length f/4 lenses for his patrols. On April 28, 2000, he detected a magnitude 10.0 object at the position of the 1917 nova in Aquila. The same night another Japanese patroller, Minoru Yamamoto, captured the object in outburst, too. Of course, the object had not been seen to recur as a nova before, so this was exciting news. CI Aquila is also an eclipsing binary: its orbital axis is inclined at virtually 90 degrees to our line of sight.

UW Per: The Biggest Mystery?

Despite major successes with many of the previously 'unobserved in outburst' objects on the ROP and similar lists, there are still some real challenges for recurrent nova or dwarf nova outburst hunters. Arguably the most puzzling object still on the list is the enigmatic UW Per (see Figure 2.26), not to be confused with UV Per, a short distance away. It took a bit of detective work on this star in the 1990s to prove that this object really did outburst in 1912, even if its precise position was slightly in error. UW Per, and its neighbor UV Per, were both discovered by Charles Robert D'Esterre (born Charles Roberts in 1876). However, although UV Per, discovered on November 13, 1911, has been verified as a UGSU dwarf nova with a 93-minute orbital period, and has been seen in outburst many times, UW Per's nature is still a mystery.

This observing section is not meant to delve too deeply into historical matters, but, undoubtedly, if UW Per is spotted in outburst again the kudos for the discoverer will be immense. So, a bit of a history lesson is, I believe, justified. UW Per was discovered on a photographic plate exposed by D'Esterre on January 9, 1912. He was using a very large Newtonian reflector for that era, one of 38-cm aperture and 1.89-m focal length (i.e. f/5). His photographic plates were somewhat insensitive by film standards of the early twenty-first century, not to mention digital detectors. They were called "Imperial Flashlight" plates, and with them he could typically reach magnitude 18 in guided 3-hour exposures, although 15.5 in 70 minutes was more typical of his results.

In 1995, after some investigative work by this author and Guy Hurst, the Royal Astronomical Society (RAS) librarian Peter Hingley unearthed the original plates of the UW Per outburst from 1912. D'Esterre had claimed an outburst magnitude of 13.5 on January 9, 1912, although Guy Hurst estimated it was actually between 14.7 and 15.1 on the original plate. D'Esterre claimed UW Per had faded to 16.5 by January 27, 1912, but Hurst's estimate made it 15.8. On further plates not found by

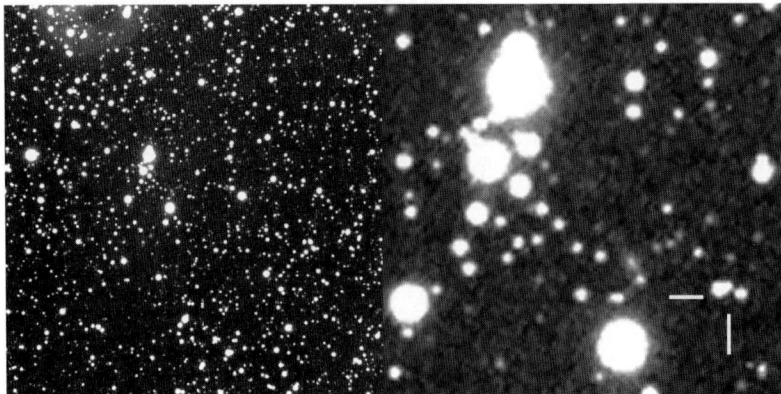

Figure 2.26. The field of the enigmatic UW Per. Since 1912 this object has never been positively recorded in outburst, although there are claims that it has been seen visually. The closest object to the 1912 photographic outburst is a star of magnitude 19.9. Its approximate position is indicated in the enlargement. Main image (left) by the author using a 0.35-m Celestron 14 at f/7.7 on February 6, 2007; the wider field is 13 arcminutes square. An SBIG ST9XE CCD was used. 2 × 180 sec exposure. North is at the top.

Hurst, D'Esterre estimated his new 'nova' was at magnitude 17.3. These plates were exposed on March 6 and 7 in 1912. This object might have been dismissed as a straightforward nova if not for some tantalizing claims to have seen UW Per visually by subsequent observers in 1915, 1917, 1919, and 1947. Specifically, Hartwig and Zinner claimed to have seen UW Per at magnitude 14.3 on August 15, 1915, and again, at 14.5, on January 27, 1917, plus 'seen clearly' on January 30, 1919. Almost 30 years later, on January 23, 1947, Himpel claimed to have seen UW Per 'brighter than mag 14'. However, in the 60 years since Himpel's claim, no further confirmed sightings have been made.

The discovery of the original plate in the RAS library at Burlington House, Piccadilly, enabled one issue to be cleared up, or did it? For the first time it was possible to check the position of the star using modern methods. The precise position of D'Esterre's suspect up to 1995 had assumed that, at minimum, it was a star that was normally recorded on modern images at somewhere between magnitudes 17.0 and 17.6 and appeared highly reddened and elongated, that is, a double star. This star was indicated as being the same object as UW Per in the classic work, *A Reference Atlas and Catalogue of Galactic Novae*, by Hilmar Duerbeck. Astrometry of this elongated double star has formed the basis of UW Per's position for decades; thus, the following positions for that star are often quoted as follows:

Duerbeck 02 h 12m 29.682s +57 05' 19.93'' (2000.0)
Downes & Shara (Old) 02 h 12m 29.59s +57 05' 19.7''
Downes & Shara (Online) 02 h 12m 29.54s +57 05' 18.5''
Brian Manning (1991) 02 h 12m 29.66s +57 05' 19.3''

In 1995 a deep WIYN (Wisconsin, Indiana, Yale, and NOAO 3.5-m telescope at Kitt Peak, Arizona) image of the UW Per field was taken that easily resolved the two stars (A and B), a third nearby star (C), and a very faint blue star (D) – 4 UW Per candidates within a 4 arcsecond radius. The astrometric positions and V magnitudes of these candidates were found to be

- Star A 02 h 12m 29.74s +57 05' 19.43'' (2000.0) mag 17.7
- Star B 02 h 12m 29.27s +57 05' 20.29'' mag 18.8
- Star C 02 h 12m 28.61s +57 0' 17.88'' mag 18.9
- Star D 02 h 12m 29.54s +57 05' 18.46'' mag 19.9

For a variety of deductive reasons the faint star D is the one currently assumed, by professional astronomers, to be UW Per in its quiet state. The online Downes and Shara catalog uses this position. This star has a blue color that is typical of dwarf nova systems (as the red dwarf's contribution is minor). The Indiana Roboscope has been monitoring the position of UW Per for more than 15 years, but the measurement aperture on the sky means that all four stars are included in the measurement. The Roboscope light-curve gives an average magnitude of 17.3 for the region, with a magnitude variation of ± 0.3 magnitudes. If the faint blue star is causing the variation in brightness it could be varying considerably, say between about magnitudes 18.3 and 21.5. Nevertheless, only one outburst has been photographically recorded, and that 1912 plate, when measured by British amateur astronomer Brian Manning (an expert astrometrist), yielded a different position of 02 h 12m 29.02s +57° 05' 15.8'', that is, 4'' west and almost 3'' south of star D's position. So maybe the real progenitor of UW Per is fainter than magnitude 22, and it is a classical nova after all. But whichever way you look at it, if UW Per outbursts again, any amateur who spots it will have bagged one of the rarest outbursting CVs of all time!

AY Lac

AY Lac was thought to be either a Mira, UG dwarf nova, or classical nova when discovered by Hoffmeister in 1928; however, it is now thought to be either a recurrent nova or a rare UGWZ type CV. It has only been seen in outburst twice, in 1928 by Hoffmeister and in 1962 by Gessner. The second outburst offers critical evidence of its recurrent nature, and, after almost half a century, the discovery of a third outburst, in the modern era, would allow much science to be carried out and the true nature to be determined. The range appears to be at least 7 magnitudes.

SV Ari

SV Ari was discovered (as Nova Ari 1905) by M. and G. Wolf on Heidelberg plates on November 5, 1905. It was reported to have brightened from magnitude 22 to magnitude 12, but the original reported position may well be in error, as the field has now been imaged by the WIYN telescope, which showed a magnitude 22 object as the most likely candidate. Tantalizingly there was a possible sighting at magnitude 15.7, on September 2, 1943, by Himpel and Jansch, but no spectra was taken, so this could be a recurrent nova and the field definitely needs monitoring.

EU Sct

EU Scuti, or Nova Scuti 1949, was discovered by C. Bertaud at the Paris Observatory on July 31, 1949. It reached magnitude 8.4 on August 5, 1949, but had faded to magnitude 11.4 only 6 weeks later. The candidate star, at minimum, appears very red and so the secondary star may be a red giant, as with the recurrent novae RS Ophiuchi and T Corona Borealis. Detection of a further outburst would confirm the classification as a recurrent nova.

FS Sct

FS Scuti was discovered by S. Arend, Observatoire de Bruxelles, on July 19, 1952. As the object was almost dead on the Scutum/Aquila border it was referred to both as Nova Scuti and as Nova Aquilae. The nova had actually peaked in brightness some 4 weeks earlier, at magnitude 10.1. It took 12 weeks to fade to magnitude 13. Professional astronomers think that subtle 'ellipsoidal variations' in the quiescent light output of FS Sct may mean a black hole is the primary component. Amateurs should certainly keep a close watch on this one.

Chapter 3

Solar Flares, Giant Prominences, and Flare Stars

Solar Flares

Our Sun is the closest star to us, by a factor of a quarter of a million. With modern equipment as inexpensive as $450, massive solar flares and prominences can be seen on its surface, changing in the space of minutes. Some solar physicists think that the next solar maximum (2010/2011) could be the most energetic ever studied, so being prepared for any major outburst is a sensible strategy.

The Sun follows a 11-year cycle of activity and, at every maximum, we see the disk sporting many large sunspots and having outbursts that, occasionally, can bring down entire national grid power systems here on Earth, as well as threatening the safety of astronauts and orbiting satellites. The most dramatic solar events are called coronal mass ejections or CMEs (see Figure 3.1). These events were not observed until the 1970s, when orbiting satellite detectors first recorded them. They are the bigger brothers of the lower-energy, but more frequent, solar flares, often witnessed by astronomers.

CMEs resemble enormous bubbles expanding out from the solar disk, caused by solar magnetic fields becoming unstable on a global scale. Billions of tons of solar material can be hurled outward at up to 2000 km per second; it then ploughs into the slower-moving solar wind, and, if the CME is heading in our direction, energetic subatomic particles can arrive at Earth a day or two later. Earth's magnetic field provides protection from the constant solar wind, which compresses Earth's magnetosphere, forming the so-called 'bow shock' shape on the sunward side. However, a small subpercentage of particles break the defenses and, if the remnants of a CME hit Earth's magnetosphere, spectacular aurorae can appear at Earth's polar regions and at lower latitudes; the resulting voltages generated in long power lines can bring down entire national grid systems. Needless to say orbiting satellites are also vulnerable to such electromagnetic onslaughts, as are ground-based radio communications. Studying the solar threat to Earth, its people, power lines, and satellites is a major part of the investigation undertaken by the flotilla of spacecraft now studying the Sun.

For hundreds of years the only part of the solar disk that could be studied from Earth was the photosphere, the blindingly bright yellow surface that is visible in heavily filtered astronomical telescopes or by safely projecting the solar image onto a piece of white card. However, in the last few decades the Sun has been studied at a

M. Mobberley, *Cataclysmic Cosmic Events and How to Observe Them*, DOI: 10.1007/978-0-387-79946-9_3, © Springer Science+Business Media, LLC 2009

Figure 3.1. In this 2002 SOHO image of a coronal mass ejection (CME) the Sun's face has been replaced by a simultaneous ultraviolet image. The solar environment out to a million kilometers beyond the Sun is shown, but the CME still extends beyond the limit of the frame. Near-solar-minimum CMEs occur every week but can occur twice a day at solar maximum. Credit: SOHO Consortium, ESA, NASA.

variety of ultraviolet, (see Figure 3.2), X-ray, and gamma-ray wavelengths by increasingly sophisticated space probes. These have increased our understanding of the processes going on in the nuclear fusion reactor at the Sun's core and in the turbulent electromagnetic world at the visible photosphere and beyond.

It is tempting to think of the yellow photosphere as being a solid surface below which the gaseous solar furnace resides. In fact, apart from the central core, we can realistically regard all of the Sun as being a huge incandescent, unbelievably hot ball of plasma, up to the photosphere, at the photosphere, and even beyond the photosphere. We see the photosphere as if it were a solid surface because it is opaque, and we are looking straight through the tenuous gas that lies above it. But the Sun does not have a solid surface at all. It is just one big ball of plasma, and the part that stretches way outside the surface-like photosphere, such as the corona, is what we humans can only enjoy during a total solar eclipse.

If only the photosphere were transparent and astronomers could glimpse the inner machinery of the Sun! In fact, that blindingly bright impenetrable top surface is only about 400-km thick. On a globe almost 1.4 million km in diameter that visible surface is, relatively speaking, thinner than the skin on an apple. Nevertheless, it is what Earth-based astronomers see, and it contains the famous 'rice-grain' granulation visible in high-resolution ground-based images.

Figure 3.2. This spectacular image in extreme ultraviolet light is a frame from a movie recorded on November 9, 2000, by the TRACE spacecraft. It shows coronal loops towering over an active solar region. The hot plasma contained in arching magnetic fields is cooling and raining back down on the solar surface. The flare material associated with the event hit Earth's magnetosphere some 30 h later. Credit: TRACE Project, NASA

The sunspots, so well known to all solar observers, are, perhaps surprisingly, regions on the solar surface where intense magnetic fields have actually reduced the energy arriving from the convective layer beneath. Sunspots are therefore considerably cooler and less bright than the rest of the visible solar surface, and their darkest parts represent the coolest and most magnetically active areas. Sunspots and the effects they generate are just like bar magnets and iron filings in those early physics lessons at school. They group together in twos, one with a positive polarity and the other with a negative polarity, usually in an east-west pairing. Magnetic arches join the two sunspots, and these solar equivalents of the schoolroom bar magnet's iron-filings manifest themselves as dramatic prominences when the Sun's rotation takes them to the solar limb.

Prominences

Prominences at the solar limb, especially when the Sun is very active, can be truly spectacular. The largest prominence ever recorded was that of June 4, 1946. The photographs of this event, recorded at the Climax Observatory in Colorado, amazed the scientific world. That prominence was known as the Grandaddy prominence, and it extended 200,000 km above the solar surface and stretched across more than half a million kilometers of the solar limb. Such prominences are extremely rare, and so the chances of a really large one being visible during the few minutes of totality at a total solar eclipse are very slim. Nevertheless, at the total solar eclipse of May 29, 1919, famous for its role in proving Einstein's theory that gravity bends light, a superb prominence was photographed by the British Eclipse Expedition led by Arthur Eddington at Principe Island in the Gulf of Guinea, off the west coast of Africa. Although not as spectacular as the awesome Grandaddy prominence of 1946 it was an incredible prominence to be seen at a total solar eclipse, and nothing larger has been seen during an eclipse since. The 1919 prominence towered some 100,000 km above the solar surface. What a sight that must have been! Large solar prominences tend to occur more when the Sun is very active and sunspot numbers are high. As we shall see shortly, the Sun has a 11-year period of activity.

Although sunspots are cooler than the surrounding photosphere they are still very hot, and as bright as an arc lamp, in real terms. Their darkness is just a relative effect. The temperature of the photosphere is typically about 5800 K compared to around 3500 K for a sunspot and a staggering 15 million K at the Sun's core.

The chromosphere (from the Greek *chromos*, meaning 'color') lies immediately above the Sun's visible photosphere. It has a depth of some 2500 km, so it is considerably thicker than the photosphere, although still only another skin on the apple. Ultra-high-resolution images at hydrogen-alpha (H-alpha) wavelengths show that the outer edge of the chromosphere, when looked at from a shallow angle, such as from the foreshortened limb view, resembles a spiky mountain range. Indeed, the spiked extensions are actually called spicules. These spicules are short-lived jets of gas (they only live for maybe 5 or 10 minutes) traveling out from the main body of the chromosphere. They can shoot up to more than 10 km in height (so the mountain peak analogy is very apt) and then fall back to the average chromosphere surface just minutes later. In the best images they resemble mountains of iron filings on a sheet of paper with a bar magnet underneath. Indeed, solar physicists think that the spicules may be involved in the conveying of magnetic fields out from the Sun to heat the next region of the solar atmosphere: the mysterious corona. This field of research is one of the hottest (pun intended) in solar physics, and almost all spacecraft currently studying the Sun have at least one instrument trying to work out how heat is conveyed from the cooler chromosphere into the million-degree-plus solar corona.

The 11-Year Cycle

Solar maxima, when the Sun is very active and has loads of sunspots on its face, occurred in 1906/1907, 1917, 1928, 1937, 1947, 1957/1958, 1968/1969, 1979/1980, 1990/1991, and 2000/2001. The period of 11 years is not all that regular, though. There was a 17-year period between the solar maxima of 1788 and 1805 and just more than 7 years between the maxima of 1829/1830 and 1837. However, in the twentieth century the most extreme variation was the 9 years from 1928 to 1937. Solar minima occur, not surprisingly, about midway between maxima.

In recent years, the so-called 'solar conveyor belt' hypothesis (a conveyor belt comprised of hot, conducting gas) has gained popularity through the work of Mausumi Dikpati of the National Center for Atmospheric Research (NCAR). In this theory the complex magnetic manifestations called sunspots, once they have decayed, have their old magnetic remnants carried away by this conveyor belt as it sweeps along just under the visible solar surface. In fact, this happens in both solar hemispheres. The theory says that these magnetic fields are then carried to the Sun's polar regions, where they drop down to about 200,000 km below the visible surface, about a third of the way to the center of the Sun.

During their journey under the solar surface the solar dynamo (for want of a better term) amplifies them, and the magnetic forces eventually emerge again at the visible surface as sunspots. One loop of the conveyor takes between 30 and 50 years to complete, roughly three or four solar cycles. So does this theory help predict how big the next solar maximum will be? According to Dikpati, and solar physicist David Hathaway of the National Space Science & Technology Center (NSSTC), the conveyor swept up sunspots very rapidly in the 1986–1996 period, that is, almost half a conveyor belt cycle ago.

Taking this one step further the magnetic energy from those amplified magnetic storms should re-emerge in the photosphere in the coming years, making the next solar maximum a very violent one, if history is anything to go by. Dikpati predicts the imminent years of the sunspot cycle will be 30–50 percent stronger than the previous one, with a burst of solar activity second only to that seen in 1958 and peaking in 2012. Hathaway predicts an earlier but equally violent solar maximum in 2011/2010, earlier due to the faster speed of the conveyor. Only time will tell if these predictions are correct, or if it will be 'back to the drawing board' and a re-examination of spacecraft results. Bearing all this in mind and, with the current boom in H-alpha viewing systems available to the amateur, investing in some suitable equipment is strongly advised.

Major Flares

The most massive solar flares can heat plasma to many millions of degrees Kelvin and accelerate electrons and protons to near-light speeds, meaning they can arrive at Earth in 15 minutes or less, affecting the ionosphere and disrupting radio communications.

Astronomers and solar physicists have an X-ray ranking system for solar flare strengths. The peak radiation measured from Earth in watts per square meter, in the wavelength band between 1 Å and 8 Å, forms the basis of the calculation. The detectors that are used to define this classification are those onboard the U.S. GOES (Geostationary Operational Environmental Satellite) spacecraft.

Trivial B-class GOES radiation levels signify X-ray radiation levels below a microwatt per square meter. C-class solar flares are rarely of consequence to us here on Earth. They range in power density from 1 to 10 microwatts per square meter. Then we come to the more significant flares. M-class solar flares are 'medium' sized, and they can cause short-duration radio blackouts near Earth's polar regions, with power densities of 10–100 microwatts per square meter. X-class solar flares are the really big events, and the category covers everything above 100 microwatts per square meter. Some of these X-class events are truly massive. Within each GOES class there is a linear scale from 1 to 9; in other words, an X2 flare is twice as powerful as an X1 flare (Table 3.1).

Prior to 2003 the two largest solar flares recorded were the X20 events (20 × 100 microwatts per square meter equals 2 milliwatts per square meter) that occurred on August 16, 1989, and April 2, 2001. However, on November 4, 2003, the GOES-12 instruments in Earth orbit detected the mother of all GOES solar flare events, emitted from sunspot region 10486. So large was the energy output that the GOES detectors red-lined at their upper limit of X28, equal to 2.8 milliwatts per square meter.

Table 3.1. Major solar flares in recent history measured on the GOES scale (since 1976)

Date	Class
November 4, 2003	~×40!!
April 2, 2001	×20.0
August 16, 1989	×20.0
October 28, 2003	×17.2
September 7, 2005	×17.0
March 6, 1989	×15.0
July 11, 1978	×15.0
April 15, 2001	×14.4
April 24, 1984	×13.0
October 19, 1989	×13.0

The March 6, 1989, event knocked out the Canadian power grid in Quebec

Research by Australian scientists David Brodrick, Steven Tingay, and Mark Wieringa, published in Volume 110 of the *Journal of Geophysical Research*, indicated a peak flux of 4 milliwatts per square meter, that is, twice the intensity of the previous most energetic flare radiation arriving at a GOES detector. Their research was based on the effect of the flare on Earth's ionosphere. In addition, as can be seen from the table, the November 4, 2003, flare occurred only one week after an X17-class solar flare on October 28, 2003. But even the X40-level flare would have paled beside the solar flare of September 1, 1859, which left evidence of its effects in the Greenland ice.

Now, you may think that with the flare occurring in 1859 there would be no 'live' evidence of this event, but there you would be wrong. The renowned observer Richard C. Carrington was projecting the solar image at the time the flare went off, and he described his observation in the Monthly Notices of the Royal Astronomical Society, Vol. 20, 13–15, 1860. His account was entitled: 'Description of a Singular Appearance seen in the Sun on September 1, 1859' and reads, in the poetic manner of those times, as follows:

> While engaged in the forenoon of Thursday, September 1, in taking my customary observation of the forms and positions of the solar spots, an appearance was witnessed which I believe to be exceedingly rare. The image of the sun's disk was, as usual with me, projected on to a plate of glass coated with distemper of a pale straw color, and at a distance and under a power which presented a picture of about 11 inches diameter. I had secured diagrams of all the groups and detached spots, and was engaged at the time in counting from the chronometer and recording the contacts of the spots with the cross-wires used in the observation, when within the area of the great north group (the size of which had previously excited great remark), two patches of intensely bright and white light broke out, in the positions indicated in fig. 1...

> ...My first impression was that by some chance a ray of light had penetrated a hole in the screen attached to the object glass, for the brilliancy was fully equal to that of direct sun-light; but by at once interrupting the current observation, and causing the image to move I saw I was an unprepared witness of a very different affair. I therefore noted down the time by the chronometer, and seeing the outburst to be very rapidly on the increase, and being somewhat flurried by the surprise, I hastily ran to call some one to witness the exhibition with me, and on returning within 60 seconds, was

mortified to find that it was already much changed and enfeebled. Very shortly afterward the last trace was gone. In this lapse of 5 minutes, the two patches of light traversed a space of about 35,000 miles.

The radiation from the September 1, 1859, solar flare created nitrates and beryllium-10 in our upper atmosphere. These deposits eventually became frozen into Greenland ice, and the amount of nitrates measured in core samples suggests that the Earth's atmosphere was hit by about 20 billion high-energy protons per square centimeter, making it the most energetic solar flare to hit the Earth in the past 500 years. In truth there were two geomagnetic storms in late August and early September 1859. They both spawned bright aurorae on August 28 and September 2, the latter event being a result of the flare witnessed by Carrington (and also by Richard Hodgson).

The results of the September 1/2 storm are recorded in national newspapers and ships logbooks of that era. Witnesses to the nocturnal auroral display described it as 'blood red' and bright enough to read a newspaper by. With the Moon being new on August 28 and the sky being free of moonlight, it must have been an awesome display, cloud gaps permitting. There were no national power grids to overload in the 1850s, but the world did have more than 100,000 miles of telegraph lines, and many were put out of action for hours at a time by the voltage induced by the cosmic storm.

How to Observe Solar Flares and Prominences

Although there are occasional observations of white light solar flares using standard solar filters, and even by the tried, trusted, and safe method of projecting the solar image onto a white card, in general to really see the Sun at work, and certainly to observe arching prominences at the solar limb, you need to observe at the wavelength of H-alpha.

H-Alpha Viewing

The first detailed descriptions of solar prominences were probably by Vassenius in 1733, although Stannyan, 27 years earlier, may have described them, too. After the eclipse of August 19, 1868, Jules Janssen in France and Norman Lockyer in England devised methods of observing solar prominences spectroscopically. Charles Young obtained the first photograph showing a prominence at the total solar eclipse of 1870. In 1924 George Ellery Hale invented the spectrohelioscope, which uses a spectroscope to scan the solar surface and produce an image at any desired wavelength. The spectrohelioscope is a fiendishly complex device involving moving slits or mirrors, such that the narrow spectroscopic slit scans the solar surface. A few are even in amateur hands.

An excellent book entitled *The Spectrohelioscope* has been written by Fredrick Veio. In Britain the well-known amateur astronomers Cdr Henry Hatfield and Brian Manning are perhaps the best known owners of such machines, although both are now in their eighties and do not observe as regularly as they once did. Building a spectrohelioscope would be a daunting prospect to most amateur astronomers, who traditionally have been content with simply observing sunspots in white light by projecting the image or using safe solar filters, but fortunately huge advances were made in producing narrow-band solar filters for amateurs throughout the 1970s, 1980s, and 1990s. In recent years there has been an upsurge in amateur astronomers who are imaging the Sun at H-alpha wavelengths, that is, at 656.28 nm. At this wavelength the limb prominences can easily be seen, even while the usually 'lethally bright' Sun is in the field. More expensive, ultra-narrow band, H-alpha filters can reveal subtle features on the disk, too.

H-alpha filters are not simply filtering huge spans of the visual spectrum of colors, like standard red, green, or blue glass filters. The bandwidth of a good H-alpha filter is, typically, less than 1 Å (0.1 nm), only one three-thousandth of the visual spectrum, and the filter production is highly complex and expensive. Originally Edwin Hirsch of the Daystar Company was the sole supplier of such narrowband filters for amateurs, but recently the Tucson-based company Coronado (www.coronadofilters.com) has been at the forefront of this technology and has developed a number of exciting products using advanced laser techniques. On Britain's Isle of Man, Solarscope (www.sciencecenter.net/solarscope/doc/about.htm) also offers quality H-alpha filters of 50-mm aperture. Both of these modern companies produce precisely tuned, ultra narrow line width classical Fabry–Perot air-spaced 'etalons' for their H-alpha filters. An etalon consists of a matched pair of ultra-fine pitch-polished and accurately figured fused silica plates. These have partially reflective and low absorption coatings for the desired transmission wavelength. To guarantee the essential fixed air space, the two etalon plates are skillfully assembled with the use of optically contacted spacers. Such filters have a very high throughput at peak resonance and a very narrow spectral transmission.

As one narrows the filter bandwidth centered on the 656.28-nm H-alpha line the prominences become more and more sharp, and fine H-alpha features on the disk emerge, too. In the 1980s the Baader Company advertised prominence telescopes in which a metal disk could be used with a telescope of a specific focal length to exactly occult the blinding solar disk. Using this method even a wide (and less expensive) 10-Å H-alpha filter would show the prominences, while the dazzling solar surface was hidden behind the metal disk.

However, by moving to expensive, narrower bandwidth filters the prominences and subtle surface chromosphere features can be viewed simultaneously. Coronado makes filters and small quality refractors optimized for use with such filters. The 2007 Coronado range consists of H-alpha telescopes from 40-mm to 90-mm aperture (ranging from $1700 to $12,000 in price) as well as individual filters priced from $900. These units typically have bandpasses less than 0.7 Å. By stacking two matched H-alpha filters together, a bandpass finer than 0.5 Å can result.

Recently, Coronado's wider, 1-Å bandpass, 40-mm aperture f/10 PST, or Personal Solar Telescope, has made H-alpha imaging affordable to many and, coupled with a webcam, spectacular pictures of prominences can now be obtained for an outlay of only $450. The PST is mainly just a prominence telescope and will show few fine details on the solar disk, but it is a remarkable price breakthrough. It may be thought that 40 mm is a very small aperture, but it is sufficient to resolve prominences only a few arcseconds in width and perfectly compatible with typical daytime seeing. Like nearly all H-alpha systems a filter 'de-tuning' collar is provided to optimize the view, and in use it gives a sort of '3D' effect, as tweaking it can enhance major disk detail or limb prominences. The reason for this 'de-tuning' is that solar flares and CMEs are amazingly fast-moving events that can mean they are Doppler shifted to be outside the narrow passband of the filter.

De-tuning moves the passband, typically by as much as ±1 Å to enable such fast-moving features not to be missed.

Although Coronado makes dedicated and safe H-alpha telescopes, many dedicated H-Alpha imagers choose not to go down this route. Why? Well, with today's webcams, ultra-short exposures, and high-speed USB 2.0 interfaces, a 90-mm aperture is simply not always enough when you get good seeing. Daytime seeing is usually very poor indeed, but sometimes it is good enough to allow instruments of 150-mm aperture and larger to reach their full potential. In addition, imagers who already own, say, a quality 100-mm refractor will be loathe to shell out thousands of dollars more than necessary when they already own the optics, but not the filter. (Coronado's 90-mm filters can easily be attached to most of the TeleVue apochromats.)

Of course, it is vital that the solar imager who constructs his or her own H-alpha system does it safely. There are dangers in making, or modifying, any solar observing system, unless you never plan to use it visually. In general there are two approaches to the H-alpha telescope system. It is not solely about manufacturing a high-quality etalon. In the first, a front aperture filter (usually with additional filters at the eye end) fits over the front of the telescope. Thus, Coronado's narrowband Solarmax filters can be fitted to, say, a TeleVue refractor, and, at the eyepiece end, within the mirror diagonal, another interference filter, namely, a Coronado blocking filter (BF), is used. Neither component can be used alone. The focal length of your system determines which BF filter (diameter in mms) you need to accommodate the whole solar image. Thus, a BF 10 is for focal lengths of up to 1000 mm (solar disk 9-mm max), a BF15 for up to 1500 mm (solar disk 13-mm max), and a BF30 for up to 3000 mm (solar disk 26-mm max).

In the second type of approach, for the observer with a larger aperture and long-focal-length telescope (see Figure 3.3), who may well want to buy H-alpha filters from the Daystar company, the main filter is at the eyepiece end, but an energy rejection filter (ERF) may also be required to reduce the invisible radiation levels so eye and filter damage cannot occur.

Figure 3.3. Dave Tyler's 6-inch (152.4 mm) f/15 refractor at Flackwell Heath UK. This is used solely for H-alpha imaging. Note the red energy rejection filter/76-mm aperture stop. Image: Dave Tyler.

The absolute minimum energy rejection required for the system allows roughly 1 part in 100,000 maximum transmission for UV wavelengths; 1 part in 1000 maximum transmission for near-IR wavelengths; and 5 parts in 1000 maximum transmission for the far IR. The ERF does not significantly attenuate the visual band, as that is where the H-alpha line resides. (For the visual part of the spectrum 5 parts in 100,000 is normally considered the borderline safety point, but, for extra safety, 1 part in 100,000 is usually adopted.) Of course, the H-alpha bandpass of less than 1 Å is so narrow anyway, literally thousands of times narrower than the visual spectrum, that no massive filtering at this precise wavelength is required for a comfortable visual view.

In practice, a high-quality 0.7-Å filter, placed in the visual band, will let through about 10 percent of the incident light within that narrow waveband. Essentially, the ERF on a system where the main H-alpha filter is at the eyepiece end is there to make sure the invisible IR and UV wavelengths do not get through to cook the filter and heat the telescope's interior. In fact, the UV (not IR) is the main area where radiation needs to be blocked, as reflected UV within an H-alpha system can heat the internal optics and degrade the filter over time. Recently Baader Planetarium has offered a new range of premium ERFs called Cool-ERF filters from 70 mm to 180 mm diameter. These filters incorporate a sophisticated coating that produces a truly cold focus by eliminating the vast majority of the ultraviolet and infrared radiation.

Daystar's filters, used by the most dedicated amateurs with apertures typically above 90 mm, come in a variety of designs incorporating a tuning T-Scanner that compensates for temperature variations from the nominal 23°C the filter is optimized for, and a heated version that keeps the filter at that temperature. Daystar filters with passbands as low as 0.3 Å can be acquired, but for optimum performance an f-ratio of f/30 or higher before the filter is necessary.

In 2006 Coronado introduced another affordable narrowband filter for amateur astronomers: a Calcium (CaK) filter for viewing the violet lines at (3933 Å and 3967 Å). Such filters show the glowing hydrogen plage regions surrounding sunspots, and the supergranulation, especially well.

H-Alpha Imaging

With the plethora of imaging devices now in the market the amateur astronomer is spoiled for choice as to what device he or she can safely use to image the Sun. In addition, even though H-alpha (and Calcium) filters are perfectly safe, you can have the added confidence that, if anything goes wrong, you will only damage the detector and not your retina.

When amateur astronomers first look through an H-alpha filter they are struck by the deep red color of the image. It is immediately obvious that this is a redder red than you see in everyday life, and the color may well be a bit off-putting at first. However, this is no problem for the CCD detector in a webcam, which is very sensitive at 656 nm. A monochrome webcam like the ATiK 1HS or the Lumenera SKYnyx 2-0 is the best choice, as it is unnecessary to use a color webcam for such narrowband work. (Figure 3.4 shows a typical arrangement). However, digital SLRs, and even small non-SLR fixed-lens digital cameras, have been successfully employed and even hand-held to the eyepiece. The Scopetronix company makes excellent digital camera to eyepiece adapters for this sort of application (www.scopetronix.com.)

Figure 3.4. Dave Tyler's high-speed Lumenera CCD camera attached to a 0.6-Å H-Alpha Daystar filter at his refractor's focuser end, enabling limb prominences and H-alpha surface details to be imaged. Image: Dave Tyler.

As we have already mentioned, Coronado's $450 Personal Solar Telescope (PST) has sparked a revolution in H-Alpha solar imaging, but the PST itself does have quite a few drawbacks where imaging is concerned. The PST's focal plane only protrudes a tiny distance from the eyepiece end, necessitating short Barlow lenses, eyepiece projection, or afocal (looking through the eyepiece with the camera lens) imaging. This focal plane restriction has forced many amateurs to dismantle their PSTs, always a very risky procedure where the Sun and eye safety is involved, unless you plan to use the device purely for imaging. Remember, the PST has built-in blocking filters to keep damaging IR and UV radiation out of the system and your eyes. Another potential problem, when imaging the whole Sun, is that it is often the case that features on one limb are optimally tuned but appear out of focus on the other limb. Such anomalies can be overcome by taking various images and combining different quadrants of the Sun into a mosaic. Some very impressive images have been taken with the humble PST and with more customized systems (see Figure 3.5).

One problem often experienced by both solar and lunar observers is that of stray light reflecting off surfaces within the optical train and ruining an otherwise great image. In many cases, where a small area of the Sun is being imaged, this is caused by unwanted filtered light from the whole solar disk. With a bit of detective work it is often found that the source of the problem is light from the whole filtered Sun entering the Barlow lens and scattering around the shiny interior of the Barlow itself. The nozzle of the CCD camera/webcam is often wide enough that this scattered light can end up hitting the CCD detector. There are two simple solutions to this problem. Firstly, the webcam/CCD nozzle can be fitted with a cap/dia-phragm such that the CCD effectively peers through a hole little bigger than itself, into the Barlow. Secondly, a black card aperture stop can be fitted to the Barlow lens to restrict the area of the Sun being accepted to just that of the region being imaged. Some trial and error experimentation is often needed in this area to determine the optimum aperture stop required.

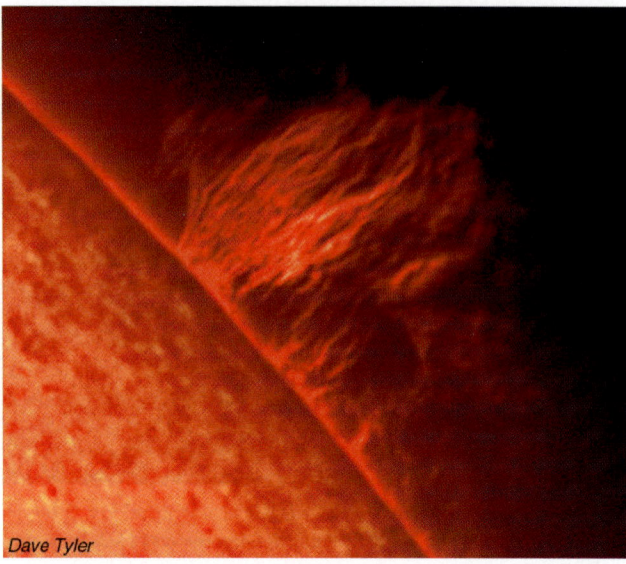

Figure 3.5. A superb limb prominence captured by Dave Tyler using the equipment shown in Figures 3.3 and 3.4. This composite is a stack of hundreds of the best frames from an imaging session on September 2, 2007. Image: Dave Tyler.

Some DSLR users have reported an interference fringe pattern effect when using an H-alpha system, with the very narrow wavelength causing an optical interference between the DSLR camera's inbuilt filter and pixel grid. To avoid such problems you can get a DSLR modified by Hutech such that the camera's IR blocking filter has been removed, or you can contact existing users of DSLRs who do not have a problem, to see which DSLR and filter arrangement works for them. The Canon 20-Da model (now discontinued) was optimized for astrophotography but still featured a filter in front of the CMOS detector; however, that filter had been modified to let through more of the IR wavelengths than normal. It should be noted that where the camera's inbuilt filter has been removed, the focus point might be slightly modified; in other words, the camera may seem to be in focus as viewed through the optical viewfinder but is fractionally out of focus in reality. Viewing the image on the camera's LCD on full zoom will reveal where the true focus point is. (Some advanced DSLRs have a ×10 magnified 'live focus' facility.)

Alternatively, you can just use monochrome webcam imaging. Webcams and video cameras have far less pixels than a DSLR, so you may wish to take wide field images by imaging loads of small regions and creating a mosaic in Photoshop or Paintshop Pro. For example, in the latter package the image canvas can be enlarged and the 'clone' brush used to copy images onto the larger canvas.

Because high-resolution imaging is so seeing-dependent, using the same techniques as the planetary webcam observer (see *Lunar and Planetary Webcam User's Guide* by this author) will secure the best images, that is, by using a monochrome webcam and processing the best stacked frames in Cor Berrevoet's Registax package. Of course, once you have the final stacked solar image there is nothing to stop you coloring it a more pleasant yellow/gold color in Photoshop or Paintshop Pro. The Sun, in H-alpha, is simply a deep red color with no other colors present. Many imagers prefer to simply represent this as a monochrome image, especially if obtained with a high-speed monochrome camera like Lumenera's SKYnyx. However, some artistic license is permitted in this field, and it seems to be popular for imagers to convert their images from monochrome to RGB and then reduce the blue channel to zero, raising the red channel value and then tweaking the green channel to tint the whole picture to the right shade of orange or red. Making the prominences fiery red in the best prominence shot and then adding the best globe image detail as a more orangey color gives the impression of a really angry and exciting Sun, rather different than the weird deep-red view seen visually through the same instrument. Solar observers often find that the limb regions appear far too dark when compared to the central disk (even in white light images) once a bit of contrast has been applied. However, this can easily be cured by a bit of tweaking with the gamma function in Paintshop Pro or Photoshop.

In general, and especially in the cheaper, wider-bandwidth filters, the solar disk is much brighter than the prominences, and each may require a different exposure or degree of image processing. Cor Berrevoet's freeware Registax is often used for stacking, sharpening, and basic processing if a webcam or high-frame-rate video camera is being used to freeze the seeing. For whole-disk images with a DSLR the best solution is to use a longer exposure setting to record perfect prominences, but with an overexposed disk and a shorter exposure setting to just expose the disk features correctly. Each image can then be optimally processed and, with your favorite software package, you can simply crop out the disk detail within the solar diameter from the shorter exposure image and paste it on top of the overexposed disk in the longer exposure prominence optimized image. If you are purely interested in limb prominences on a tiny section of the limb, the overexposed disk in the longer exposures can be digitally masked with a black 'fill' command in Paintshop Pro or Photoshop.

If you are using a digital camera you may well find that different color channels in the RGB image show different amounts of contrast, even though the image is supposed to be purely red and the green and blue channels should be black. This is because digital camera CCD or CMOS green and blue pixel filters have some red leakage and, remarkably, the contrast in the green and blue channels can sometimes be more useful than the red (i. e., if there is a moiré patterning artifact). With apertures of 100 mm or so some truly spectacular images of even modest solar prominences can be obtained.

Finally, whatever type of solar observing you do, do it safely. If in any doubt, let the camera and the webcam do the imaging and not your eye. Cameras and webcams can be replaced, retinas cannot.

VLF Radio Detection of Solar Flares

Even the most dedicated solar flare enthusiasts cannot be expected to stare at a projected image of the Sun for every minute that it is above the horizon, or even stare at a PC screen relaying webcam images. This strategy would not tell you whether the flare particles were going to hit Earth either, although it would tell you what was happening on the Sun. Of course, professional solar observatories, such as those at Boulder in Colorado monitor the Sun all the time it is above the horizon, and a host of solar astronomy satellites are looking at our nearest star at X-ray and UV wavelengths, too. Nevertheless, it is good to know that there is a simple method available to amateurs who possess a bit of electronics knowledge; this method enables them to detect and record solar flare particles entering Earth's atmosphere, and it involves monitoring very low frequency (VLF) radio transmissions.

At high altitudes in Earth's atmosphere the air is so thin that free electrons can exist for short periods before being recaptured by ions, which are positively charged atoms. This region of the atmosphere is known, not surprisingly, as the ionosphere. There are enough free electrons in the ionosphere to affect the transmission of low-frequency radio signals. The ionosphere layers that are important in this regard are called the D layer and the E layer. The D layer extends from about 50- to 90-km altitude and the E layer from about 90- to 120-km altitude.

The VLF radio transmissions that are most usefully affected by solar flares ionizing the D and E layers are typically those operating at between 15 kHz and 30 kHz. The VLF bands are usually defined as being between 3 kHz and 30 kHz, but the frequencies below 9 kHz are not allocated by the International Telecommunication Union and so can be used without a license.

In practice radio stations transmitting at around 20 kHz provide the most useful sources for monitoring the state of the ionosphere. The D layer is only present during the daytime and is ionized directly by solar radiation. Not surprisingly it is highly sensitive to solar flares impacting the atmosphere. VLF signals tend to be guided by the D layer acting as a ceiling, at least in the daytime. At night the higher-altitude E layer tends to reflect higher-frequency signals that can increase in strength during the night hours. The UV and X-rays arriving due to solar flares increase the daytime ionization of the D layer, causing sudden ionospheric disturbances (SIDs) that can dramatically alter the signal strength of VLF radio transmission. If a PC (or a chart recorder) is set up to monitor VLF signal strength when a solar flare hits the atmosphere, the resulting trace will be dramatic.

In the case of a violent X-Class flare the VLF signal strength usually dips violently and then bounces back to an abnormally high strength. This pattern may well be repeated with another dip occurring in the next hour or two. The rise and fall is far more dramatic than can be caused by any natural phenomena. Of course, the choice of which 10–30-kHz radio station you tune into to monitor its signal strength will depend on your precise location on Earth, and some trial and error will be necessary.

For many years the American Association of Variable Star Observers (AAVSO) has operated a SID program, and they can supply advice on making a SID detector, including plans and circuit boards for assembling an antenna, tuner, amplifier, analogue to digital converter, and data logger. Their members Art Stokes, Cap Hossfield, and Joseph Lawrence have provided the electronics expertise and more information is available at www.aavso.org/observing/programs/solar/SID MonOverview.shtml.

Of course, technology moves along quickly these days, and the options for monitoring and, especially, recording SIDs have increased dramatically in recent years, with data loggers capable of interfacing with PC USB and Ethernet ports now becoming available. The British Astronomical Association (BAA) has recently designed a 'Plug and Play' Observatory design for monitoring SIDs and solar microwave bursts at 2.695 GHz (a protected frequency) and feeding the data into a PC. For more details on this see www.britastro.org/radio/

Flare Stars

On the Sun, solar flares are thought to be caused by the quick release of energy during magnetic reconnection events. The visible solar surface is laced with incredibly powerful and constantly changing magnetic fields, and the sunspots are regions on this surface where intense magnetic fields have actually reduced the energy arriving from the convective layer beneath. When the magnetic field switches to a lower-energy state in these regions the excess energy can be dumped into the plasma in and around the magnetic field. The plasma can actually be accelerated to relativistic speeds in these circumstances and radiates greatly in the ultraviolet and X-ray regions of the electromagnetic spectrum, producing a solar flare. Even gamma rays can be emitted in such situations, and in extreme cases white-light solar flares are observable in the visual spectrum.

A similar set of events are thought to occur in so-called flare stars, but because they are rather faint stars normally, especially in the blue end of the spectrum, their flares can drastically increase the magnitude of these stars. Typically, a flare star within amateur observing range might rise by a magnitude in the visual band while rising 5 magnitudes in the ultraviolet. In contrast, even the most violent flares on our Sun have little effect on its overall brightness. Another factor is that the flare stars' flares themselves can be significant compared to the size of the star, maybe covering 20 percent of the star's circumference!

Flare stars are known to be dim red dwarf stars, although some may be brown dwarfs. Because they contain only a fraction of the Sun's mass (typically between 0.1 and 0.6 solar masses) and have a low luminosity, the best known examples within amateur range are all within 60 light-years of Earth. Flare stars are often referred to as UV Ceti stars. The star UV Ceti (UV by name and by nature) was originally called Luyten 726-8, after W.J. Luyten who noted variable-brightness hydrogen emission lines and variable spectra in the stars V1396 Cyg and AT Mic in the 1920s. In the next couple of decades similar behavior was observed in V371 Ori, WX UMa, YZ CMi, and DO Cep. But it was only when Luyten 726-8 was studied by Joy and Humason at Mt Wilson, who observed brief flares of the star of almost 50-fold in brightness, with a return to a normal state in under 24 hours, that the name UV Ceti was allotted and a new field of variable star research was opened up. Flare stars are quite cool in quiescence, around 2500–4000 K, but during a flare they can rise to 10,000 K. Many are in binary systems, and some are also members of the class of stars known as BY Draconis stars, or 'spotted variables.'

How to Observe Flare Stars

Flare stars are difficult to observe simply because the flares themselves are unpredictable. Unlike with variable stars that vary slowly, over days and weeks, a flare star will, typically, outburst by 1, 2, or 3 magnitudes in the space of a few minutes and then fade back to its typical (quiescent) level in half an hour or several hours. Many of the keenest variable star observers like to observe as many stars as possible during a night. After a while they find they can memorize the star fields for each variable star, and so making a visual magnitude estimate for each star is a very short affair, lasting less than a minute. Then they will move onto another star to see if anything odd is happening. By checking stars this rapidly there is more chance of finding the star that is in outburst, or the star that has faded, that particular night.

Statistically, only a small percentage of variable stars will be showing any dramatic change from night to night, so covering a large number (100 is quite typical for the sharpest observers) pays dividends. However, this strategy does not work with flare stars, so this field of research needs a special type of observer, one that is not averse to long periods of study of a particular star. Simply checking a flare star once every clear night is unlikely to catch it in outburst. Another strategy might be to observe many variable stars per night, but keep coming back, every 10 minutes or so, to the flare star. Of course, modern CCD equipment is well-suited to long, tedious, sad, and mind-numbingly boring monitoring tasks like this! It is easy to set up a CCD camera to take an image every few minutes, throughout the night, and no problem to check the images the next day for any signs of an outburst.

Despite the frustrating aspects of visual flare star observing, there is a certain mysterious appeal to watching a star so close to our own Solar System outbursting so violently. Although CCD detectors are very quantum efficient, their sensitivity at the blue end of the spectrum is quite poor compared to their red and infrared sensitivity. When a standard U or B filter is used to monitor a flare star a loss of 2 or more magnitudes is quite common. When submitting a report of a flare star's magnitude during an observation period it is sensible to record an estimate every few minutes when the star is in outburst, as the flare magnitude can change rapidly. Much smaller flares than can be detected visually can be bagged with a CCD, so precision photometry is a useful project. Variations of 0.1 magnitudes or less are very difficult for the human eye and brain to detect, and, if the star is on the limit of your equipment, 0.2-magnitude variations can be a major challenge, too. One of the big attractions of visual flare star observing is that outbursts are so unpredictable and observers so rarely all stare at the same star that if you do see a flare you may be the only person on Earth observing it visually.

Top Flare Star Targets

Because of the low absolute magnitudes of these dim red dwarfs not many lie within easy visual range of amateur astronomers and their telescopes. Nevertheless, there are half a dozen examples that are prime targets, and these are listed below. All these stars lie within 50 light-years of Earth, and all but one are within 20 light-years, so when observing them, we are witnessing their behavior in the very recent past, within our own lifetimes even, unlike with so many other energetic astronomical events.

One consequence of these stars being so nearby is that their proper motion, that is, the rate they drift across the sky relative to the more distant stars, is high, and can be measured by amateur astronomers on images taken less than a year apart. Thus the R.A. and Dec. positions are only approximate. Typically these stars drift at 2 or 3 arcseconds per year. In the sample chosen here the star WX UMa zips along at 4.5 arcseconds per year, whereas V371 Orionis, at 49 light-years distant, moves at the slowest rate, that is, only 0.2 arcseconds per year. We have also entered Barnard's star into the table (Table 3.2), despite only one major flare having been witnessed. This nearby star is a record breaker, streaking along at 10 arcseconds per year. A few figures of the fields of these stars appear in this chapter and we have provided the rest as 5 arcminute thumbnails in the Resources section of this book.

UV Ceti is the top-priority flare star to monitor, but being at almost Dec. −18 it is never high up for European observers, even in the autumn. It can outburst to magnitude 6.8 in extreme cases, largely due to its very close proximity to our Solar System. At a distance of only 8.7 light-years it is only fractionally further from us than the brightest star in our night sky, Sirius. It is the sixth closest stellar system

Table 3.2. Nearby flare stars and their characteristics. The approximate absolute magnitude is that of the star in quiescence. These stars drift measurably across the sky even over years, so the R. A. and Dec positions will alter but are not enough to move them outside an eyepiece field within the lifetime of the early twenty-first century reader of this book!

Variable	Alt. name	Approximate coordinates	Magnitude	Ab. Mag.	Distance (light-years)
UV Ceti	Luyten 726–8	1 h 38m 51 s – 17°57.5'	6.8–13.0	~16	8.7
YZ CMi	Ross 882	7 h 44m 43 s + 03°34.1'	11.8–13.2	~14	19.4
EV Lac	BD + 43 4305	22 h 46m 52 s + 44°20.3'	9.5–11.5	~13	16.5
AD Leo	BD + 20 2465	10 h 19m 30 s + 19°52.4'	9.4–10.9	~12	16.0
V2500 Oph	Barnard's star[1]	17 h 57m 48 s + 04°41.6'	~9.6	~13	6.0
V371 Ori	HIP 26081	5 h 33m 46 s + 01°57.2'	10.0–11.7	~11	49.0
WX UMa	Lalande 21258	11 h 05m 53 s + 43°30.6'	13.0–14.8	~16	15.8

[1]Barnard's star has been included primarily because of the flare of July 17, 1998, which suggested that this very nearby star was worth monitoring. It is a rewarding object to image anyway, as two exposures only months apart can clearly show its motion even with an amateur telescope

to us, and only the triple star Proxima/Alpha Centauri system, Barnard's star, Wolf 359, Lalande 21185, and the Sirius binary system are closer.

Proxima, Barnard, and Wolf are also classified as small red dwarf flare stars, but their outbursts are not in the same category as the other six stars (excluding Barnard) listed in the table. UV Ceti is itself a double star, but the flare star in question is designated as component B. Component A (BL Ceti) is listed as a flare star, too, but its variability is, at best, trivial in comparison. The two stars orbit their common center of gravity every 26.5 years at a separation of 2.3 billion km or roughly 2 light-hours (a few arcseconds when viewed from our distance). This is less than the distance of Uranus from our Sun (mean distance 2.9 billion km). Although UV Ceti can vary between magnitudes 13 and 6.8, a typical outburst sees it flare from around magnitude 12 to 10, typically every 10 hours or so. Being at a declination of almost –18 degrees in Cetus the Whale, UV Ceti is not well placed for far northern observers. However, it is reasonably placed in the month of October and can be found in the same field of view as the galaxy NGC 648, which is just 8 arcminutes to the north of the flare star. The star can be back to its normal quiescent state in 10 or 15 minutes after a flare.

YZ CMi can be found just 2 degrees to the southeast of the brilliant star Procyon, which is impossible to miss in the northern hemisphere during winter months. At ultraviolet wavelengths it can flare by 4 magnitudes during an outburst. In February 1983 YZ CMi flared in the visual, UV, and radio bands, and that outburst was particularly well studied.

As well as being one of the brightest flare stars in the sky, EV Lac is observable for most of the year from North America and northern Europe. Technically, it never sets from latitudes above 45 degrees north, but in practice it is simply well placed at some time of night from these latitudes in all but the spring months. An outburst monitored by the British observer John Saxton is shown and described in Figure 3.6.

AD Leo is situated some 5 arcminutes west-northwest of second-magnitude Algieba (Figure 3.7) in Leo's Sickle. This makes the field easy to find but a bit painful to stare at for long periods with such a bright naked-eye star in the same

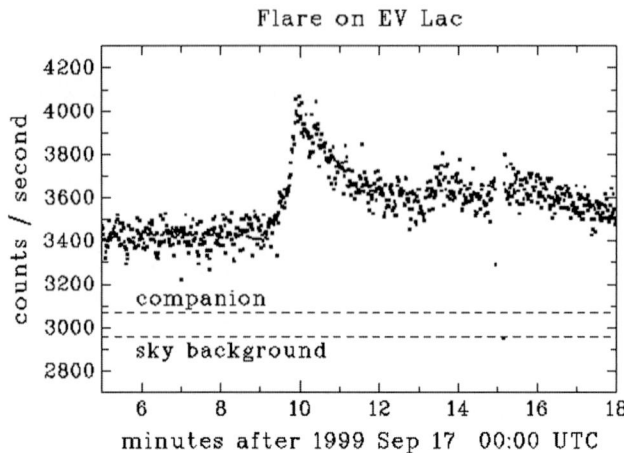

Figure 3.6. A flare of EV Lac recorded by John Saxton, Great Britain. His telescope then was a 21-cm Newtonian, and the old photometer he was using employed an IP21 family Photomultiplier tube and DC amplifier. The following comments are John's:

"On 1999 September 15–16, I monitored EV Lac for 70 minutes with a B filter without detecting any flares. On 1999 September 16–17, I observed the star again. After an hour, I was rewarded with the spectacular flare shown in the figure. I was away from the telescope during the rise part of the flare, but on returning, I was surprised to find the signal much higher than usual. I looked into the guiding eyepiece, and EV Lac did appear brighter than it had done several minutes previously. That said, I might add that I was very tired by that time, and would not relish the prospect of trying to detect flares visually. The figure shows the measurements binned to 1 second time resolution.

EV Lac brightened by almost a magnitude in 1 minute, with most of the brightening in about 20 seconds. There was then a slower fading, followed by a long 'tail' during which the brightness returned to normal around 20 minutes after the flare had started (not all the tail is shown in the figure). The gap in the lightcurve at 00:15 UTC is a guiding correction.

We can estimate the energy radiated by the flare by calculating the area under the light curve. According to Leto et al, the quiescent luminosity of EV Lac is 3.49×10^{22} watts in B. This flare radiated about 1.5×10^{25} Joules in B alone. Leto et al show the energy distribution of flares they observed: only 10% radiate 1×10^{25} Joules or greater in B. For comparison about 1×10^{25} Joules are radiated in large solar flares, about a quarter of this appearing at visible wavelengths. My EV Lac flare was thus an order of magnitude larger than a large solar flare."

Light-curve and all comments by kind permission of John Saxton.

eyepiece field of view. (Figure 3.8) Ideally a high-power eyepiece should be used and the telescope positioned so Algieba is just outside the field. A four-night intensive study of AD Leo in March 2000 by the University of Colorado revealed eight major flares during that period.

Barnard's star, also known as V2500 Ophiuchi, has generally been regarded as a flare star, but it is not. However, at least one major flare has been detected, and it is a fascinating object in its own right. Arguably the last of this extraordinary observer's truly memorable discoveries was that of the motion of this 9th-magnitude star in Ophiuchus. In May 1916 Edward Emerson Barnard spotted what appeared to be three variable stars in a straight line on photographs he had exposed in 1894, 1904, and 1916. He realized it was the same star but had moved 4 arcminutes south in the intervening 22 years. Today that star is called Barnard's

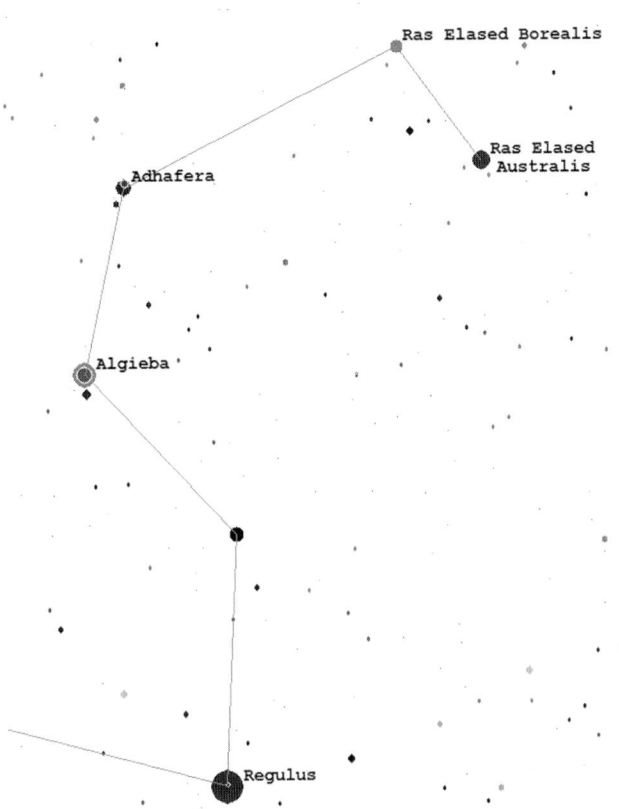

Figure 3.7. The star Algieba is second only in brightness to Regulus in the Sickle of Leo and lies in the same telescope field as AD Leo.

Figure 3.8. The flare star AD Leo is indicated by markers in this 15 × 15 arcminute field, which contains the bright star Algieba (see previous figure). North is up. Image: STScI DSS.

star, or V2500 Ophiuchi. Barnard had found the fastest-moving star in the sky, because of his obsessive astrophotography and the star's small distance from us of only 6 light-years. On July 17, 1998, Diane Paulson and her colleagues at NASA's Goddard Space Flight Center in Greenbelt, Maryland, reported a major flare from Barnard's star while they were taking high-resolution spectra with the observatory's 2.7-meter telescope. Their aim was to detect planets orbiting the nearby star, but instead they detected the flare. When analyzed some time later it appeared that the star's hot blue flare was at a temperature of 8000 K, more than double the cool star's typical temperature of 3100 K. It was estimated that the flare would have been half a magnitude or more visually and lasted for at least an hour. Paulson was reported as saying this would be a great star for amateurs to observe for flares. Unfortunately, in the last 10 years, few amateurs have taken up this challenge, but maybe a reminder is necessary; hence, we have included the star here.

V371 Orionis is another star that is situated close to a much brighter and well-known object, but in this case, it is far enough away to avoid the brighter star dazzling the observer. V371 is just 2 degrees north of Mintaka, (Figure 3.9) the westernmost star of the three in Orion's belt. This flare star is the only one described in this section that is more than 20 light-years from Earth. At 49 light-years it is far more distant and intrinsically the brightest star in its quiescent state; in other words, it has the most luminous absolute magnitude. But it still only shines with roughly 1/300th of the luminosity of our own Sun. Figure 3.10 shows the quarter-degree field.

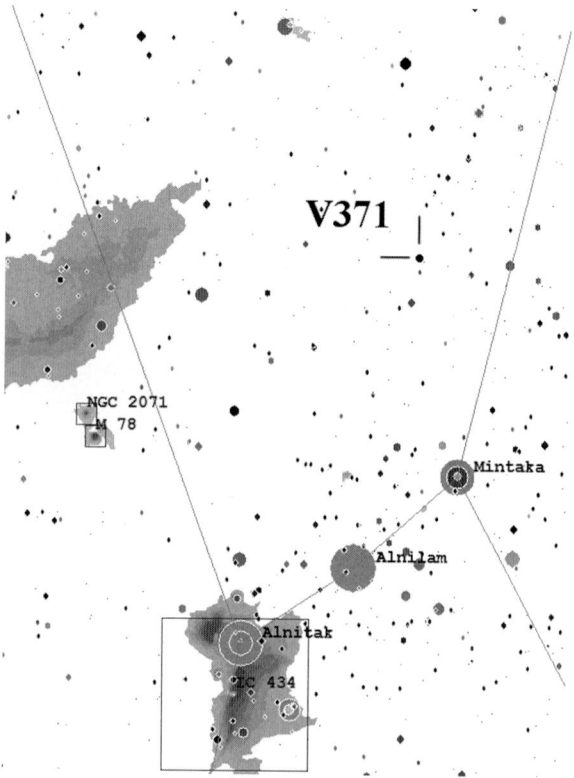

Figure 3.9. V371 Orionis is in a very familiar part of the northern hemisphere winter night sky, just above the Orion's Belt star Mintaka.

Figure 3.10. The flare star V371 Ori is indicated by markers in this 15 × 15 arcminute field, which is just above the Orion's belt star Mintaka (see previous figure). North is up. Image: STScI DSS.

Figure 3.11. Flare stars are among the very nearest stars to us, and therefore they can have a significant movement (called proper motion) across the night sky. These images of the motion of the flare star WX UMa in the intervening 41 years between the first (right hand) and second (left hand) Palomar Sky Surveys show it to have moved more than 3 arcminutes west (and slightly north), that is, a proper motion of more than 5 arcseconds per year. The star is a binary, and both components can be seen to have moved in this image. The variable is the fainter component, that is, WX UMa B. Image: STScI DSS.

The final star described here is WX UMa and, like EV Lac, it is another circumpolar star at a very similar declination. For North American and northern European observers the star is only really badly placed in August and September when, at midnight, it is low down near the northern horizon. Like UV Ceti the star has two components, but, in this case, the much brighter (magnitude 8.7) component A, 31 arcseconds away, greatly outshines the flare star component B. At an outburst magnitude of 13 this flare star is primarily a target just for the dedicated, large telescope using, flare star observer. This double-star system is sometimes cataloged as Gliese 412. Its proximity and motion relative to us mean it moves rapidly across the night sky over periods of years and decades. See Figure 3.11.

Chapter 4

Bright Supernovae and Hypernovae

How to observe supernovae has been covered exhaustively in the author's book, *Supernovae, and How to Observe Them*. Nevertheless, when talking about the highest-energy events amateur astronomers can observe, we must include supernovae, though the subject will not be covered in as much detail here.

Supernovae mark the end state of a star's life. In the case of a massive star the progenitor collapses into a neutron star or a black hole when it runs out of fuel. The other category of supernova involves a smaller star, but, perhaps surprisingly, an even brighter explosion. When a white dwarf star goes supernova, it is totally blown apart by a massive thermonuclear detonation (see Figure 4.1). All that is left is its companion star, which was feeding it the hydrogen that led to the explosion.

The newcomer to supernovae might imagine that the situation is therefore pretty clearcut, that there are big star supernovae, small star supernovae, and that is the subject totally wrapped up. No such luck! Every supernova is different, and supernovae are classified by their hydrogen content, not their size. The careful reader will have noticed that this situation, where a companion star feeds hydrogen to the exploding white dwarf star, is remarkably similar to the scenario for less violent cataclysmic variable stars – novae, recurrent novae, and dwarf novae. As we have seen, in the case of dwarf novae, flare-ups on the accretion disk surrounding the white dwarf are the mechanism, whereas in the case of novae and recurrent novae a thermonuclear explosion on the white dwarf's surface is the cause. In both cases the steady stream of hydrogen from the companion star is the root cause.

Type Ia Supernovae

In the case of those supernovae where the progenitor is a white dwarf, the stream of hydrogen from the companion star is still the source of the outburst. However, in this extreme case the white dwarf has a very specific mass of around 1.4 solar masses (one solar mass being the mass of our own Sun). Such supernovae are classified as Type Ia.

What is the significance of 1.4 solar masses? Well, as dealt with in Chapter 1, a white dwarf is a very dense object, with the mass of a star compressed into the size of an Earth, and a density of roughly 1 ton (1000 kg) per cubic centimeter. Every teaspoonful of a white dwarf would weigh about the same as a small family car, if it was transported to Earth in the same state. On the surface of the white dwarf, due to the star's powerful gravity, the weight would be five orders of magnitude higher! This bizarre compressed star would like to contract even further, into a neutron star, but it is prevented from doing so by fast-moving electrons and a quantum effect called 'The Pauli Exclusion Principle.' The pressure

M. Mobberley, *Cataclysmic Cosmic Events and How to Observe Them*,
DOI: 10.1007/978-0-387-79946-9_4, © Springer Science+Business Media, LLC 2009

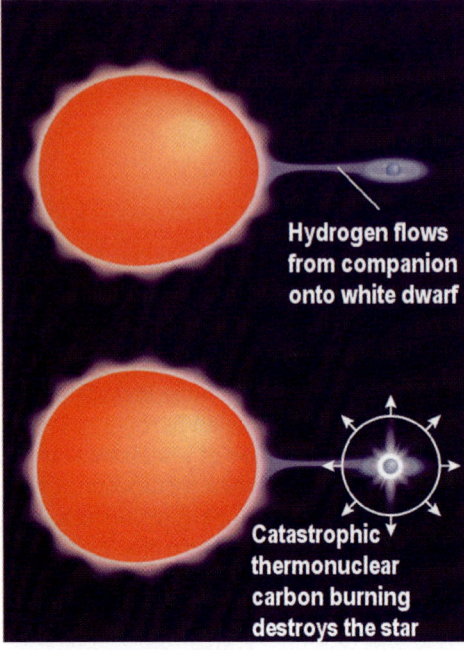

Hydrogen flows
from companion
onto white dwarf

Catastrophic
thermonuclear
carbon burning
destroys the star

Figure 4.1. In a Type Ia supernova cata-strophic thermonuclear carbon burning destroys the star.

exerted by the electrons is called degeneracy, and the material in a white dwarf is said to be electron degenerate. The dividing line at 1.4 solar masses is often referred to as the Chandrasekhar limit. For a white dwarf more massive than 1.4 solar masses the forces of gravity will overwhelm the electron pressure, and the white dwarf will collapse to a neutron star. At 1.4 solar masses the star is teetering on the brink. When hydrogen from the companion flows onto the surface of the electron-degenerate white dwarf it turns from hydrogen to helium and adds to the mass of the white dwarf. At a crucial point the white dwarf starts to collapse, which heats the carbon in its core and triggers a colossal outpouring of energy that blows the white dwarf apart. The technical term here is 'catastrophic thermonuclear carbon burning.' The star never makes it into the neutron star phase and is blown to smithereens before it can exceed 1.4 solar masses.

There are other Type Ia scenarios, too. One involves the coalescing of both binary stars, causing a similar outcome. A third scenario involves the accretion process triggering runaway helium burning just beneath the white dwarf's surface. An asymmetric explosion then triggers the star's destruction.

Massive Progenitor/Core-Collapse Supernovae

In the normal life of a single star that has used up all its hydrogen fuel and collapses, the 1.4-solar-mass limit is equally important. Beyond the Chandrasekhar limit electron pressure is not enough to prevent further gravitational collapse, and a star will collapse into a neutron star; essentially, just one big solar-mass-sized atomic nucleus maybe 30 km in diameter. Once more the Pauli exclusion principle steps in to prevent further collapse, but applied to neutron pressure and neutron degeneracy this time. The density has now jumped by more than a hundred-million fold compared to the density of a white dwarf, from 1000 kg per cubic centimeter to more than 100 billion kg per cubic centimeter, that is, approaching the density of an atomic nucleus, except one that is held together by gravity and not the strong nuclear force.

Some astronomers think that neutron stars have masses of between about 1.4 and 2.5 solar masses, and that beyond that (at the tongue-twisting Tolman–Oppenheimer–Volkoff limit) a further collapse into a quark star is possible (quarks being the building blocks of neutrons). Either way, a neutron star or a quark star both have unbelievable densities and escape velocities that are significant fractions of the speed of light. The reader will have guessed what is coming next. Yes, that's right, above about 5 solar masses the collapsing star's gravity overwhelms any nuclear force, and the escape velocity exceeds the speed of light. The star disappears into a black hole of its own making. This is what happens in non-Type Ia supernovae. These are simply massive stars that have run out of fuel. Without the thermonuclear reactions to support the outer layers the star eventually collapses into a neutron star or black hole. So Type Ia and non-Type Ia supernovae have something in common, namely a superdense object at the root of the problem. In Type Ia's it is the source of the problem, along with the stream of hydrogen from the companion star. In non-Type Ia's the superdense object marks the end state of a massive star's life, whereas in Type Ia's there is no end state, just a badly battered companion star that was a witness to the white dwarf's destruction.

Let us have a look in detail now at what happens just before a supermassive star collapses causing a non-Type Ia supernova explosion. Non-Type Ia supernovae are thought to originate from the collapse of stars that were once heavier than about 8 solar masses (see Figure 4.2). When such stars swell up in size in the later stages of their development much of the unburned hydrogen can be lost, thus reducing the final mass considerably. However, with 8–100 solar masses to start

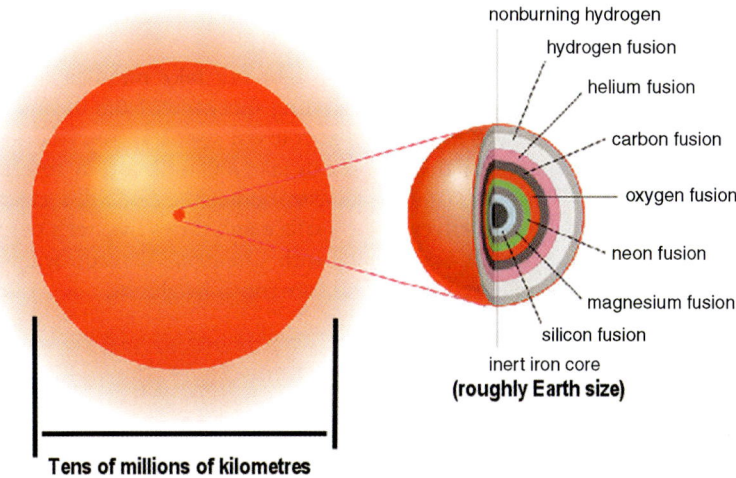

nonburning hydrogen
hydrogen fusion
helium fusion
carbon fusion
oxygen fusion
neon fusion
magnesium fusion
silicon fusion
inert iron core
(roughly Earth size)

Tens of millions of kilometres

Figure 4.2. When a giant star finally runs out of fuel, 'burning' all the elements up to iron, a core-collapse supernova results. There are various different types, but they are all non-Type Ia supernovae.

with there will always be a lot of mass still left at the end. In such stars heavier elements continue to be burnt (in nuclear reactions) as the lighter elements are used up. Thus hydrogen is the fuel source initially, then helium, then carbon, neon, oxygen, and even silicon.

Other heavy elements are produced in this unbelievably hot furnace, and each time a heavier element burns the length of time that it fuels the star is reduced. Thus carbon may keep the process going for 1000 years, neon for 1 year, and silicon for a few days. When only iron is left you get less energy out than you put in trying to get that element to fuse, and so there is an enormous gravitational attraction inside the star but no nuclear pressure to keep it inflated, as it were. The iron core, containing elements such as nickel and cobalt, will probably be at around 8 billion degrees and at a density of 500 metric tons per cubic centimeter at this point.

Exactly what happens when the pressure inside the star can no longer counteract gravity is still a source of some debate, but things happen very quickly. Within a split second of the core collapse the iron core's protons and electrons are crushed into a neutron star state and an unimaginable number of neutrinos are released. As we have already seen, a stellar core of that size may settle for being a neutron star, or if there is more than 2 or 3 solar masses left in that 20- or 30-km-diameter core it may become a quark star (subject to the theory being accepted) or beyond about 5 solar masses go on to become a fully fledged black hole.

Before we get engrossed in the fate of the star's core, we need to remember that there were potentially millions of kilometers of the rest of the star surrounding it when the core itself collapsed. As the core collapsed the lighter elements surrounding it got drawn down, too, but once they hit the solid (and here we really do mean 'solid') 20-km-diameter atomic nucleus that is the neutron star core they have no choice but to rebound. It is this rebounding shock wave that rips the rest of the star apart and allows us to see a supernova explosion.

As far as the finer details of such events are concerned, well, there is still much debate among supernova experts. At the heart of massive stars enormous convection currents and magnetic fields play a big part, as (probably) do the neutrinos. Neutrinos, in a normal environment, do not interact with normal matter; but the environment of a collapsing star is hardly normal. The neutrinos may convey as much as 10^{46} J worth of energy away from the core!

Massive stars are large objects, and even light can take minutes to traverse a distance equal to their diameters. The rebounding shockwave from the infalling star bouncing off the solid nucleus is traveling at colossal speed by everyday standards (a few percent of the speed of light), but it can take half an hour, or several hours, to reach the surface of the star. To a casual astronomer the star will seem perfectly normal until that shockwave arrives at the surface and flings the star's outer layers into space. Only if the observer is equipped with a neutrino detector will he or she know that the core has collapsed before the supernova brightens.

Supernova Subtypes

Now let us just have a recap at this point. There are two basic types of supernovae: Type Ia, where a white dwarf of 1.4 solar masses is blown to smithereens by catastrophic nuclear burning, and non-Type Ia, where a massive, or supermassive star, runs out of fuel and collapses.

The reader who is new to supernovae might, at this point, assume that they would be classified according to whether they were of exploding white dwarf type or collapsing massive star type. That would seem logical, but, sadly, this is not how the classification system works. Type I supernovae are ones that are hydrogen-free (according to their spectra) and Type II supernovae are ones that contain hydrogen and look vaguely sunlike in this regard. Although the Type Ia's are the special and superluminous exploding white dwarfs, the rest are all massive stars. The 'rest' comprise Types Ib, Ic, II, IIb, IIn, II-L, and II-P (see Figure 4.3 for the complete set of options!).

The Type Ia supernovae, caused by white dwarf detonation, are, undoubtedly, the most highly prized supernovae, because they are the brightest and because they enable astronomers to measure the distance to the supernova. This latter feature is possible only because Type Ia's are all supposed to be around 1.4 solar masses, and so they all explode with a similar luminosity. Like most things in astronomy, though, this rule is not set in stone. Type Ia's do come in different luminosities, but astronomers can adjust for this by observing how quickly they fade and by studying their spectra.

In recent years the presence of 'dark energy' has largely been deduced by studying very remote supernovae well beyond amateur range. Type Ia supernovae typically peak at an absolute magnitude of between –19 and –20; in other words, that would be their brightness at a distance of 10 parsecs, or 32.6 lightyears. Our own Sun has an absolute magnitude of +4.8, and so a Type Ia supernova outshines it by a factor of roughly 24 magnitudes, or 4 billion times. Put another way, a Type Ia supernova 1 light-year from Earth would look as bright as our Sun, 8.3 light-minutes away, at least, for a few weeks while it was near its peak luminosity. When a Type Ia supernova occurs in a small galaxy it can outshine the entire galaxy when viewed from Earth. When one considers that the progenitor star was a relatively feeble white dwarf the outburst is all the more incredible.

By comparison, the non-Type Ia massive progenitor supernovae typically peak at an absolute magnitude of –17: that is, they are 10 times fainter, but still hundreds

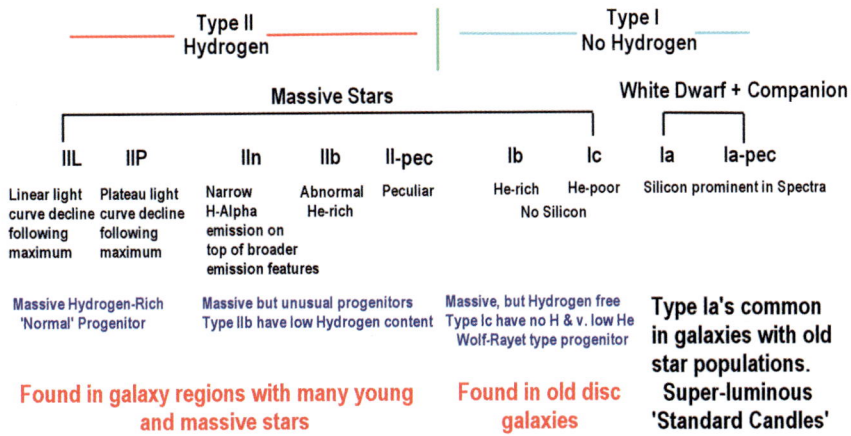

Figure 4.3. The complex classification system of supernovae.

of millions of times brighter than our Sun, while at their peak. In addition, the magnitude increase is less impressive, too, typically only about 10 magnitudes, as these huge stars were cosmic searchlights even before their cores collapsed. For this reason it has proved possible to locate a few of these huge progenitor stars, before they exploded, in archival photographs and images.

Let us now briefly look at the different types of massive progenitor supernovae, that is, Types Ib, Ic, II, IIb, IIn, II-L, and II-P. Remember, Type Is are hydrogen-free and Type IIs have hydrogen. The white dwarf Type Ia's may not show hydrogen in their spectra, but they do show evidence of silicon, sulfur, and magnesium. Conversely the big, core-collapse but hydrogen-free supernovae of Types Ib and Ic do not exhibit silicon lines, but do show oxygen and magnesium. Type Ibs show helium lines as well. Type II supernovae all reveal hydrogen in their spectra, as does our own Sun, of course. Thus, this type of spectrum is often called the 'solar abundance.' The various subcategories of Type IIs can be summarized as follows:

IIb: Helium rich; much hydrogen removed by tidal winds.

IIn: Narrow H-alpha emission lines on top of broader emission features. Slow final decline.

II-L: Type II with a linearly fading light-curve after maximum.

II-P: Type II with a standstill plateau feature shortly after the decline from maximum.

II pec: Peculiar Type II spectrum.

A few images of recent supernovae are shown in Figures 4.4, 4.5, and 4.6.

Figure 4.4. Supernova 2003ie in NGC 4051. A Type II supernova discovered by Ron Arbour. Imaged by the author on September 23, 2003, with a 0.35-m f/7.7 Celestron 14 and SBIG ST9XE. 3 × 120 seconds.

Figure 4.5. Supernova 2006bp in NGC 3593. A Type II supernova discovered by Koichi Itagaki. Imaged by the author on April 27, 2006, with a 0.35-m f/7.7 Celestron 14 and SBIG ST9XE. 3 × 240 seconds.

Figure 4.6. Supernova 2007av in NGC 3279. A Type II supernova discovered by Ron Arbour. Imaged by the author on April 7, 2007, with a 0.35-m f/7.7 Celestron 14 and SBIG ST9XE. 2 × 120 seconds.

The 'Champagne' Supernova

Despite what we have said about supernova types above, sometimes a supernova comes along that really surprises astronomers and makes them wonder if they need to change their ideas. Supernova 2003 fg (originally called SNLS-03D3bb) was discovered on April 24, 2003, as a faint 20th-magnitude speck in an anonymous galaxy in Bootes, close to the border with Ursa Major and not that far from the bright star Alkaid. The supernova seemed like a standard Type Ia until astronomers studied the spectrum and light-curve in detail. They concluded that the white dwarf progenitor was 2 solar masses, thereby smashing the 1.4-solar-mass Chandrasekhar limit that defines the point at which Type Ia's detonate and sets them up as standard brightness candles. The very size of the universe is these days based on the measurement of Type Ia supernovae, and so one that does not obey the rules threatens to shake the whole of cosmology.

After some consideration astronomers held their view that SN 2003 fg was a 2-solar-mass Type Ia supernova but that it was an exceedingly rare event. Explanations for how such a heavy white dwarf supernova could occur included the possibility that it was caused by the merger of two smaller white dwarfs, or that a fast-spinning white dwarf could help to counteract the overwhelming force of gravity. Either way, such a rare supernova can only help us to have a new understanding of these violent explosions, a discovery event worthy of a champagne celebration, and a term first coined by the British group Oasis with their song of the same name in 1996. In 2007 another Type Ia supernova appeared that seemed to have a 'super-Chandrasekhar-mass'. SN 2007if was discovered with the 0.45-m ROTSE-IIIb telescope at McDonald Observatory at a magnitude of 19.5. The object rose to magnitude 17 and had a redshift of 0.07, implying an absolute magnitude brighter than –20. The spectra appeared similar to the Champage supernova of 4 years earlier, implying a second 'super-Chandrasekhar-mass' Type Ia supernova had, indeed, occurred.

The Brightest Supernovae

Supernovae appear to be very rare events in the Milky Way. None have been seen in our own galaxy for 400 years, even though, statistically, we might expect one every 30–50 years. On the basis of the number of active pulsar remnants in our galaxy (an estimated one million), an even higher supernova rate of one every 10 years can be predicted. Yet this is simply not what we see.

The occurrence rate for the five best-documented naked-eye supernovae, occurring between 1006 and 1604, hints that maybe one a century is more typical for our galaxy and that the current 400-year dearth is just a statistical quirk. If we remind ourselves that Type Ia supernovae peak at an average absolute magnitude of –19.5, and the rest peak at about –17, and then feed in our galaxy's diameter of 100,000 light-years (we are slightly more than half way from the center to the edge), we can judge the brightness spread of various supernovae from our viewpoint. Without any intervening matter we would expect that supernovae occurring in our own Milky Way galaxy would always appear at visual magnitudes above zero, even for the furthest possible examples of Type II. Of course, tens of thousands of light-years of intervening dust in the galactic plane may hide these

Table 4.1. The brightest naked-eye supernovae for which convincing evidence exists. Being naked-eye objects these were all within our own galaxy, except for the most recent supernova, 1987A, which was in the Large Magellanic Cloud, a satellite galaxy of our own

Year	Mag.	Distance (light-years)	Constellation	Remnant/Progenitor	Comments
1987	2.5	170,000	Dorado/LMC	Sanduleak −69° 202a	Best studied SN
1604	−3	13,000	Ophiuchus	G004.5+06.8	Kepler's SN
1572	−5	15,000	Cassiopeia	3C 10	Tycho's SN
1181	0	9000	Cassiopeia	3C 58	
1054	−5	7000	Taurus	Crab Nebula	Messier 1
1006	−9.5	3000	Lupus	PKS 1459-41	
393	−3	N/A	Scorpio	RX 51713.7-3946 ?	
386	N/A	15,000	Sagittarius	G11.2-0.3	
185	−2	7500	Centaurus	G135.4-2.3 (probably)	Bright comet?

from view, but, even taking this into account, the current 400-year drought is puzzling, and an average of one naked-eye supernova per century just prior to that is far less than theory predicts. Our ancestors noticed brilliant events in the sky well before 1006, and there are, puzzlingly, few accounts of these, either. Only three supernovae prior to 1006 – in A.D. 393, A.D. 386, and A.D. 185 – come with reasonably convincing credentials. Table 4.1 details the best known historical and nearby supernovae.

In fact, in addition to the supernovae in Table 4.1, there is another that seems to have avoided visual detection in the late seventeenth century. The youngest known supernova remnant in our galaxy and the strongest radio source in the sky (beyond the solar system) is called Cas A, as it resides in Cassiopeia at a distance of some 11,000 light-years. Professional images of the expanding shell indicate that the supernova responsible must have detonated around the year 1667, some 350 years ago. Interestingly, the British astronomer John Flamsteed cataloged a star near the position of Cas A on August 16, 1680, and called it '3 Cassiopeiae.' But there is no star in that position today. So Flamsteed may have been the only observer to note, or record, this supernova's position. As it was not noted by other observers it could hardly have been very bright. Nevertheless, it does appear that in the late 1600s a Milky Way supernova did occur but was largely missed. Maybe our galaxy produces far more of these dust-obscured events and really does meet its theoretical quota? Fortunately, in the twenty-first century we no longer have to miss any supernovae. With the number of astronomy satellites and even ground-based detectors now monitoring space at every conceivable wavelength it is unlikely that any future supernovae in our galaxy will be missed when their radiation arrives on Earth.

Extraordinary Supernovae

By any normal definition all supernovae are extraordinary. However, from an observer's perspective, amateur or professional, a few stand out as truly exceptional and stick in the memory of all who witnessed them. Since 1885 only 13 supernovae, all outside the Milky Way, were discovered at magnitude 11.5 or brighter. There may well have been others, but supernova patrolling was in its infancy in the early decades of the twentieth century. Supernovae this bright are generally Type Ia's in galaxies within 50 million light-years, or core-collapse types within 15 million light-years. Extend the magnitude to 13 and there are more than 50 examples because we are trawling in objects twice as distant. In modern times the supernova in the Large Magellanic Cloud (LMC) had no equals.

Supernova 1987A

February 23, 1987, has gone down in astronomical history as a date of colossal importance for supernova science. On that date the brightest supernova in our skies since 1604 appeared in the LMC, a companion galaxy of our own Milky Way. This was not the long-awaited galactic supernova, but at only 170,000 light-years away it was the next best thing.

Having a supernova this close to us opened a number of scientific gateways. SN 1987A was close enough for neutrinos from the core collapse to be registered in three neutrino detectors: one in Japan, one in Russia, and one in the US. As we saw earlier, core-collapse supernovae release colossal amounts of neutrinos; however, even 10^{57} neutrinos (a popular figure in this regard) can be tricky to spot when you are 170,000 light-years away. Only (a relative term!) trillions of neutrinos will pass through every square meter by the time they reach the underground detectors on Earth some 170,000 years later, and as only one in a trillion neutrinos (depending on their energy) will react with atoms in the detectors' liquid tanks a supernova has to be within half a million light-years or so to have any chance of detection. A supernova like that occurring in AD 1006, 60 times closer, would have triggered 3,600 times the number of neutrino detections.

As it is, SN 1987A triggered 25 neutrino detections in three tanks across the world within a 15-second window, namely, at the Kamiokande II detector in Japan (12 hits), the IMB detector in Fairport, Ohio (8 hits), and the Baksan detector under Mount Andyrchi in the northern Caucusus mountains (5 hits). The neutrinos were detected at 07:36 UT on February 23, signaling the precise time of core collapse as viewed from Earth. From the subsequent visual observations of Albert Jones and the photographic observations of Robert McNaught it appears that the supernova erupted visually between 2 and 3 h later, in good agreement with supernova theory. Because the supernova was so close, the progenitor star, Sanduleak −69° 202, was a bright 12th-magnitude star and had been photographed many times prior to going supernova. Astronomers were rather surprised to find that the progenitor was a blue supergiant; they had been expecting a red supergiant. In fact, after some theorizing, they concluded that the star had been a red supergiant up to a few thousand years before its core collapsed (a few thousand years is like yesterday in stellar terms), but it had shrunk and heated up, into a 'B3I' blue supergiant before going supernova.

Being a mere 170,000 light-years away has also meant that the passage of the shockwave through the surrounding LMC environs could be tracked using the Hubble Space Telescope (HST) in the subsequent 20 years. Above Earth's atmosphere, and with a 2.4-m mirror, HST has a resolution of 2 light-weeks at the LMC's distance.

Images of the region at this resolution have provided much data, but more questions than answers. In particular, the volume of space around SN 1987A appears to be occupied by a mysterious hourglass-like figure, the waist of which is some two-thirds of a light-year from the supernova and the upper and lower rings of which are some four times further away. The waist may simply be old debris from the original star having an outburst in the distant past, some 20,000 years before it went supernova, with that debris being illuminated by the flash of the supernova two-thirds of a year later. The rest of the hourglass is much harder to explain but may be due to the star having had a black hole or neutron star companion whose jets 'painted' a pattern on the local volume of space. Ten years after SN 1987A detonated astronomers announced they had detected the supernova's blast wave interacting with that existing ring of debris, two-thirds of a light-year from the position of the progenitor Sanduleak –69° 202. Astronomers have searched for a neutron star remnant of the supernova, but so far one has not been found. Either the neutron star is still enshrouded in dense dust clouds, or the supernova actually left a black hole remnant.

Supernova 1987A had an unusual double-humped maximum in its light-curve. Those who have studied supernovae as they brighten and fade will tell you that no two supernovae are the same.

Supernova 1993J

Although northern hemisphere astronomers missed out with SN 1987A, they got lucky 6 years later with a supernova that was as far north as the former spectacle was south: both were 69 degrees from the celestial equator. SN 1993J occurred in the nearby Messier galaxy M81. Admittedly, at a peak magnitude of 10.5 on March 31, SN 1993J was 8 magnitudes inferior to SN 1987A. However, it was still the seventh brightest supernova seen since the last galactic supernova of 1604 (if we ignore the mysterious seventeenth-century visual counterpart of the radio source Cas A).

In case the reader is wondering, the other six were, in brightness order: SN1987A (obviously); SN 1885A (M31); SN 1895B (NGC 5253); SN 1937C (IC 4182); SN 1972E (NGC 5253); SN 1954A (NGC 4214). The 1885 supernova in M31 peaked at magnitude 5.8. The remaining four were 8th- and 9th-magnitude objects.

SN 1993J was discovered by a northern hemisphere amateur astronomer, a Spaniard named Francisco Garcia Diaz, who was using a compact 25-cm f/3.9 Newtonian telescope at 111×. A supernova in a Messier galaxy is extraordinary enough, but one peaking at 10th magnitude was unprecedented for northern European and North American observers. With a decent aperture you could stare directly at this supernova and really feel like you were witnessing something incredible. Conversely, most supernovae are only visible by using averted vision i.e., looking to one side of the object so its light falls on the most sensitive part of the retina (see Chapter 7). Once again another bright supernova would surprise us when the spectrum was studied. SN 1993J was originally classified as a standard Type II core-collapse supernova; that is, it had hydrogen lines in its spectrum. But within a few weeks of its discovery the hydrogen lines had become weaker, and helium lines had appeared. SN 1993J appeared to be a Type Ib core-collapse supernova with a dense helium core. After the supernova reached its initial mag 10.5 peak it plunged back to mag 11.8 a week later and then peaked again 2 weeks after that. It then commenced a steady decline at a rate of one magnitude every 7 weeks. The light-curve and spectra do not easily fit into any category of supernova, but the theorists think SN 1993J had a much lower hydrogen content than normal, possibly due to a companion star stealing the progenitor's atmosphere. This might explain the double peak and the spectrum. Once again, a progenitor star was identified, this time a 20th-magnitude candidate in one of M81's outer spiral arms.

Supernova 1980K

The sixth supernova to occur in the ultraproductive galaxy NGC 6946 flared up in the era when photographic film was still king, for both amateurs and professionals. In those days it was largely a question of whether you preferred grainy Tri-X or T-Max 400, versus hypersensitized Kodak 2415. The 3-magnitude advantage of charge coupled devices (CCDs) was not even a dream in the minds of the world's leading amateurs in 1980. Although not quite as bright as SN 1993J would be 13 years later, SN 1980K was a supernova that inspired and excited many amateur astronomers when observing supernovae was a very rare pastime, even among deep-sky enthusiasts, and discovering supernovae was little more than a fantasy outside professional circles. At a declination of +60°, SN 1980K was superbly placed for many British, European, and North Americans and, from England, it was circumpolar, too, so the object could be studied for a very long period without the frustration of it hiding behind the Sun for several months. Amateur astronomers such as Britain's Brian Manning and Alan Young enabled its decline to be followed right down until it was 18th magnitude, even without the availability of electronic detectors.

Supernova 2004dj

Supernova 2004dj, in NGC 2403, was a recent supernova that was well studied by many amateur astronomers. It was discovered on July 31, 2004, by Koichi Itagaki of Teppo-cho, Yamagata in Japan, his 10th supernova in 3 years of successful searching. At the time of writing Itagaki has discovered 31 supernovae using a large, 60-cm Newtonian. At discovery SN 2004dj was estimated to be magnitude 11.2, making it the joint eighth brightest extragalactic supernova ever observed. This brightness was not due to the supernova type (it was not a Type Ia but a Type II-P) but due to the galaxy's proximity. Sometimes classed as a member of the M81 group of galaxies, NGC 2403 is the second brightest galaxy in that group, despite not having a Messier designation. At a distance of 12 million light-years, any supernova in NGC 2403 was always going to be a bright one.

Although 2004dj was rather poorly placed at discovery (low in the north, in the northern hemisphere summer) because the galaxy was at such a high (+65°) declination, it was visible all year round from far northern latitudes. In addition, it was 160 arcseconds east (and 10″ north) of the galaxy core, so was easy to separate from the bright inner region of NGC 2403. As the supernova fell in brightness and became much better placed, amateurs were able to follow its decline, without a break, for 6 months, until it was well below magnitude 15 (see Figure 4.7).

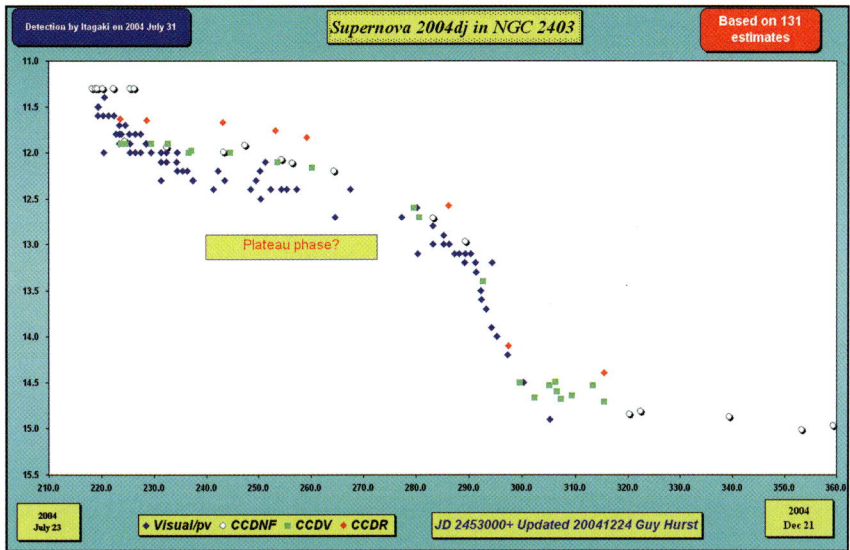

Figure 4.7. One of the best-observed supernovae in recent years was the very bright, 11th-magnitude, Type II-P SN 2004dj in NGC 2403. It was observed continuously by amateur astronomers as it declined from 11th to 15th magnitude from July to December 2003. Note the characteristic plateau phase after the initial decline. Light-curve by kind permission of Guy Hurst/*The Astronomer*.

Ultrabright Supernovae and Hypernovae

The term 'hypernova' is one that is often used to describe abnormally bright or abnormally massive supernovae, although at first glance the name itself could equally well refer to abnormally bright novae. The term is not associated with any specific supernova type, but, more often, with supermassive core-collapse supernovae (more than 100 solar masses), superluminous supernovae (that are brighter than would be expected from their presumed distance), and supernovae associated with long gamma-ray bursts (GRBs). In the latter case the core of the giant star collapses straight into a black hole, and two high-energy plasma jets emerge from the polar axes at close to the speed of light. A number of distant supernovae have been unequivocally associated with GRBs and thus qualify as hypernovae. Thus SN 1998bw and GRB 980425 (see Chapter 6) are regarded as being the optical and gamma-ray parts of the same event. Two extraordinary supernovae, classed as hypernovae, which were within amateur observation range, are detailed below.

SN 2002ap: A Messier Galaxy Supernova

SN 2002ap was discovered on January 29, 2002, by the Japanese amateur astronomer Yoji Hirose. The discovery magnitude was only 14.5, but the supernova climbed slowly to reach magnitude 12.5 some 2 weeks later. At the host galaxy M74 s distance of approximately 30 million light-years this was not overly bright; the absolute magnitude peaked at magnitude –17, typical of a core-collapse supernova. However, the spectrum revealed very broad emission features characteristic of a high-kinetic-energy outburst and a supernova of Type Ic. The spectra of the larger hypernovae 1997ef and 1998bw were similar. Astronomers concluded that the progenitor star may have had a mass of about 25 solar masses, quite small by hypernova standards but heavy enough to collapse into a black hole. The supernova may have been close to the lower mass end of the hypernova scale, making it a valuable test case for hypernova theories and a major success for amateur astronomy discoveries. It was a great supernova to follow fading, too, as its angular elongation from the center of M74 (NGC 628) was 4 arcminutes west and 2 arcminutes south, and the galaxy was well placed in the evening sky in the first few months of 2002.

Just to emphasize how strange the discovery statistics for supernovae can turn out, SN 2002ap was the very first supernova discovered in that bright face-on Messier galaxy. Yet, less than 17 months later, on June 12, 2003, Bob Evans of Australia discovered a second in M74, namely SN 2003gd. Although that second object was not a hypernova, it was also well observed, especially from the southern hemisphere.

A word or two about SN 2002ap's discoverer might be of interest here. Yoji Hirose is well known in Japanese amateur astronomy circles, and since discovering 2002ap he has discovered two more supernovae, namely 2002bo in NGC 3190 and 2005W in NGC 691. He had been searching for 22 years with five different telescopes when he found the M74 hypernova, but then discovered his second less than 6 weeks later and his third less than 3 years after that. 2002ap was found with a 25-cm f/10 Schmidt–Cassegrain telescope working at f/6.3 and mounted on a Takahashi EM 200 mounting. An SBIG ST9 CCD detector was used for imaging. It is worth noting that the M74 hypernova was discovered the day after a full Moon! With CCDs no time of the month is ruled out for supernova hunting.

SN 2006gy: The Most Luminous Supernova Ever?

On September 18, 2006, a bright 15th-magnitude supernova was discovered in the galaxy NGC 1260 in Perseus, less than 2 degrees to the east-northeast of the famous 2nd-magnitude eclipsing binary star Algol. The discovery, by Quimby, was made with the Texas Supernova Search Rotse IIIb instrument, a 0.45-m aperture f/1.95 modified Cassegrain with a 1.85-degree wide field at the 28 mm × 28 mm (2048 × 2048 pixel) detector. Months later it was deduced that the progenitor star's core had probably collapsed around August 20, making this a very slow-rising supernova indeed. The supernova's brightness finally peaked around October 25, at magnitude 14.2. Allowing for the 240 million light-year distance to NGC 1260, this gave the supernova an absolute magnitude brighter than –20; when allowing for substantial galactic reddening this brightness increased to –22. As we have already seen, most core-collapse supernovae have absolute magnitudes of –17, and even the brightest Type Ia white dwarf supernovae rarely peak above –20. So SN 2006gy was an incredibly luminous supernova. If it had occurred within 3 light-years of Earth it would have easily outshone our Sun, just 8.3 light-minutes away! Astronomers estimate the total energy output of the supernova explosion may have been as high as 10^{53} ergs, that is, 10^{46} J.

So why was SN 2006gy so violent? The best guess is that the progenitor star was truly massive, probably around 150 solar masses. Even by core-collapse supernova standards it would have been a huge star. A star this big – specifically, between 130 solar masses and 250 solar masses – may explode by a completely different mechanism to the standard core-collapse. The core of the star can produce super high-energy gamma rays with a greater energy value than the rest mass of an electron. These can interact with the nuclei in the star's atomic structure, producing particle and antiparticle pairs. This process can itself stimulate even more gamma-ray emission, creating a runaway reaction. More matter–antimatter particles are produced, and the star eventually explodes completely, rather than collapsing in on itself. The resulting supernova is a monster, even by supernova standards, and not even a black hole remnant is left. Astronomers cannot be sure that SN 2006gy was a so-called 'pair instability' supernova, but from the absolute magnitude and the slow rise to maximum brightness, it was a very unusual and superluminous event that almost certainly came from a supermassive star. In the skies of the southern hemisphere the giant star Eta Carinae may well end its life in a similar manner. That star, within our own Milky Way galaxy, is currently a 4th-magnitude object, visible to the naked eye. It lies roughly 7500 light-years from Earth. In the eighteenth and nineteenth centuries Eta Carinae brightened twice, quite substantially, peaking at a magnitude of –0.8 in 1843. These brightenings are indicative of a highly unstable star that probably has a mass in excess of 100 solar masses. Astronomers think it will probably turn into a supernova very soon in cosmological terms, which, in practice, means in the next million years. When it does, if it were as luminous as SN 2006gy, it would shine at roughly magnitude –10, or more than 100 times the brightness of the planet Venus in our night-time sky.

SN 2006jc: The Supernova that went off *twice*!

Actually, it did not go off as a supernova twice, but it is a unique event in modern times – as far as we know. Supernova 2006jc was discovered by the Japanese supernova patroller Koichi Itagaki, in UGC 4904, on October 9, 2006. From the spectra astronomers deduced that it was a hydrogen-poor supernova with strong, but narrow, helium lines superimposed on a broader spectrum typical of a Type Ic supernova, that is, a massive 'collapsar.' But when it went off astronomers recalled a supernova 'false alarm' for this galaxy that had occurred in mid-October 2004. The 2004 event was considerably fainter and disappeared after a few days, but – and here is the amazing part – it was in exactly the same position in the galaxy. After mulling this mystery over, astronomers have concluded that this huge star must have had a giant outburst 2 years prior to going supernova. This is not totally unknown, even within our own galaxy. The star Eta Carinae experienced such an outburst in 1843 (see the previous paragraph) and is another colossus. It is quite possible that the progenitor star of SN 2006jc had a mass of up to 100 solar masses, and it had a particularly odd composition, too: a star rich in helium, which tends to support the pre-outburst theory. Nevertheless the event (an outburst, followed 2 years later by a supernova) seems to be unique, but with astronomers patrolling the skies constantly in the twenty-first century, maybe another example will be discovered in the near future?

How to Discover and Observe Supernovae

Supernova Light-Curves

Without access to a large-aperture telescope and a spectroscope it might be thought that little science could be carried out from an amateur's backyard observatory. However, many amateurs are variable star observers, and, with sufficient aperture, the techniques for studying variable stars can be applied to monitoring the fading light of a bright supernova. Although not as powerful as spectroscopy, an enormous amount of information can be gleaned from the shape of a supernova's light-curve. Chapter 7 deals in detail with the various technical aspects of photometry.

At first sight the light-curves of most supernovae look virtually identical. Following a rapid rise to maximum brightness and a short stay at the peak, a gradual decline to obscurity takes place. However, look more carefully, and subtle differences emerge (see Figure 4.8). Type Ia (white dwarf progenitor) supernovae show the sharpest rise to maximum and frequently stay within a magnitude of maximum brightness for less than a month. Roughly 5 weeks after the peak a Type Ia supernova may have slumped in brightness by 3 magnitudes. Then the decline suddenly slows down, to a rate of around a magnitude dropped every 8 weeks.

At the same distance Type II-L supernovae are only a tenth as bright as Type Ia's at their peak, but the basic shape of the light-curve is fairly similar. Either side of the peak the rise and fall is a bit more sedate, but the initial decline rate persists for twice as long as with Type Ia's, that is, for about 10 weeks. At the end of this decline phase, a Type II-L supernova will typically be 4 magnitudes below the peak. The supernova then fades rather more slowly, at about 1 magnitude every 3 months.

Type II-P supernovae are very obvious from their light-curves, which are characterized by the distinctive 'plateau' phase. The rise is usually significantly slower than for Type Ia's or Type II-L's. The plateau phase commences after about 1 month and just more than a magnitude below the peak. Typically the light-curve is then horizontal for 2 months. How can this happen? Well, it is thought to be linked to a change in transparency in the star's outer layers, which are heated to 100,000 degrees by the explosion. The hydrogen is ionized, becoming opaque, and so we cannot see the decline in brightness of the inner fireball, only the constant light output from the outermost parts of the supernova. When this stage ends there

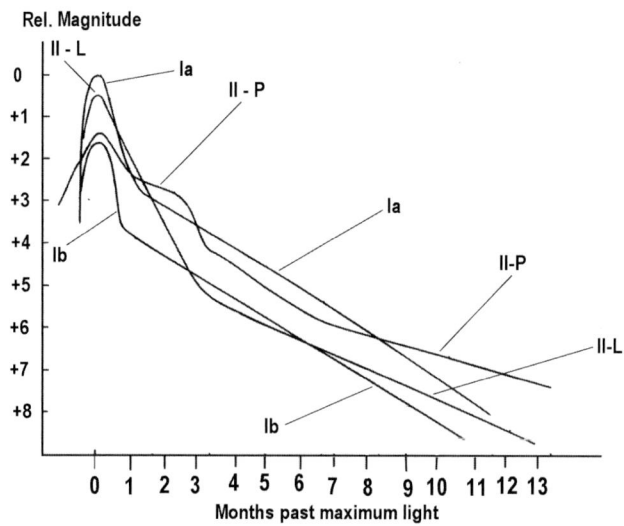

Figure 4.8. The typical light-curves for different types of supernovae. See text for details.

is usually a week-long plunge, followed by a slower decline of around a magnitude and a half over the next 3 months. The next 3-month period shows an even slower decline, typically half as fast. Type II-P supernovae are, like II-Ls, fainter than the ultrabright Type Ia's; however, there is considerable variation as big core-collapse stars come in a variety of sizes, from big to truly colossal.

Type Ia supernovae are most common in galaxies with old star populations. As we have seen, a white dwarf in a binary star system is the progenitor and not a collapsing giant star. Type Ib supernovae are rare animals and seem to occur mainly in old disk galaxies. Large young stars that have lost their hydrogen are thought to be responsible. The so-called Wolf–Rayet-type stars are, again, a likely contender. Type Ic supernovae are very rare objects indeed, containing no hydrogen and no helium, either! Their progenitor stars are probably similar to those of Type Ib. Type II supernovae – all those with hydrogen in the spectra – are common in galaxies/galaxy regions containing a lot of massive young stars.

How to Search for Supernovae

Many amateurs are far more attracted by the thought of searching for supernovae than actually monitoring their photometric decline after someone else has discovered them. A supernova discovery carries a lot of kudos, especially if it is in a bright Messier galaxy. So what are the prospects? Supernova hunting is a numbers game. Some nearby face-on spiral galaxies have produced more than their fair share of supernovae and can give false hope to the budding patroller. For example, the high declination galaxy NGC 6946 has spawned 9 supernovae between 1917 and 2008. The longest gaps between subsequent supernovae in this galaxy (since 1917) were both 22 years, namely, between 1917 and 1939 and between 1980 and 2002. The shortest gap was just 1 year, between 1968 and 1969 (Table 4.2).

Basically, NGC 6946 produces a visible supernova roughly once per decade. Compare that to the Andromeda Galaxy M31, which has not (as far as we know) produced any visible supernovae since S Andromedae reached magnitude 5.8 in 1885, more than 120 years ago. The same argument could be made of our own Milky Way Galaxy, which produced its last easily visible supernova in 1604. Of the 40 Messier galaxies (39 if we discount M102 = NGC 5866, which has produced zero supernovae), there are still 15 bright Messier galaxies that have yet to produce any visible supernovae (see Table 4.3 for the 24 supernova producing Messier galaxies).

Table 4.2. The nine supernovae in NGC 6946 up to Feb. 2008. The offset values are arcseconds from the galaxy nucleus

Designation	Date	Discoverer	Disc. Mag.	Offset	Type
2008S	Feb. 1	Arbour	17.6	53W 196S	II
2004et	Sep. 27	Moretti	12.8	250E 120S	II
2002hh	Oct. 31	LOTOSS	16.5	61 W 114S	II
1980 k	Oct. 28	Wild	11.4	280E 166S	II-L
1969P	Dec. 11	Rosino	13.9	5 W 180S	N/A
1968D	Feb. 29	Wild/Dunlap	13.5	45E 20 N	II
1948B	Jul. 6	Mayall	14.9	222E 60 N	II
1939C	Jul. 17	Zwicky	13.0	215 W 24 N	N/A
1917A	Jul. 19	Ritchey	14.6	37 W 105S	N/A

Table 4.3. Messier galaxy supernovae up to the end of 2007

Galaxy	Supernovae	Amateur Messier discoveries
M31	1885A	
M49	1969Q (not proven to be a supernova)	
M51	1994I; 2005cs	Puckett et al. (94I); Kloehr (05cs)
M58	1988A; 1989 M	Ikeya and Evans (88A)
M59	1939B	
M60	2004 W	
M61	1926A; 1961I; 1964F; 1999gn; 2006ov	Itagaki (06ov)
M63	1971I	
M66	1973R; 1989B; 1997bs	Evans (89B)
M74	2002ap; 2003gd	Hirose (02ap); Evans (03gd)
M81	1993 J	Garcia
M82	1986D (probably not a SN); 2004am	
M83	1923A; 1945B; 1950B; 1957D; 1968L; 1983 N	Bennett (68L); Evans (83 N)
M84	1957B; 1980I; 1991bg	Romano (57B); Kushida (91bg)
M85	1960R	
M87	1919A	
M88	1999 cl	
M96	1998bu	Villi
M99	1967H; 1972Q; 1986I	
M100	1901B; 1914A; 1959E; 1979C; 2006X	Johnson (79C); Suzuki (06X)
M101	1909A; 1951H; 1970G	
M106	1981 K	
M108	1969B	
M109	1956A	

If SN 1969Q in M49 is discounted (it was never proven), this number of Messier galaxy 'failures' rises to 16. Considering how well these galaxies have been patrolled since the end of World War II, and especially since the 1980s, it is perhaps reasonable to assume that supernovae in the Messier galaxies would, in most cases, only be lost when the galaxy was at solar conjunction, that is, too close to the Sun in the sky to be seen.

Up to the end of 2007, of the 39 genuine Messier galaxies, 23 (or 24, depending on whether SN 1996Q was real) have produced a total of 46 (or 47) supernovae. The top seven performers are M83 (6 SNe), M61 (5), M100 (4), M66 (3), M84 (3), M99 (3), and M101 (3). In addition, M51, M58, and M74 have each produced two supernovae since 1988.

If we look at the tally of supernova discoveries since 1983, the year in which the remarkable Australian visual observer Bob Evans discovered his first Messier galaxy supernova, and at the dawn of serious amateur supernova patrolling, we find that in the period 1983–2007 there were 19 Messier galaxy supernovae, in a 24-year timespan. Given that there are 39 Messier galaxies, this gives us an average of one supernova discovered, per Messier galaxy, every 50 years or so.

Another factor to bear in mind is that the best producers tend to be big, face-on spirals. Size is important in this regard. The more stars in a galaxy, the more chances of a supernova going off. In addition, with an edge-on galaxy, intervening dust can easily obscure our view of a supernova. Of course, there are exceptions to any rule, and Messier 33 (NGC 598) is probably the best example. Spanning 70 × 40 arcminutes and only 3 million light-years away, it would produce a superb supernova of between 5th and 8th magnitude, and, if it were a core-collapse event, the progenitor star could, almost certainly, be identified.

Yet no supernova in M33 has ever been found. Nevertheless, the average Messier galaxy statistics are very encouraging, and barely a year goes by without another Messier supernova occurring. M33 is a much smaller galaxy than M31 with, maybe, only one-tenth of the number of stars, so the rarity of M31 supernovae since 1885 is much more puzzling. In addition there are many other NGC galaxies within amateur patrol range that were overlooked by Messier and yet are just as bright and easy to patrol. The 35 galaxies listed in Patrick Moore's popular 'Caldwell' catalog are prime targets, especially as many modern telescope hand controllers provide one-touch Caldwell 'Go To' enabling easy slewing to these objects.

The beginner to supernova patrolling might imagine that with supernovae cropping up every 50 years or so in bright, relatively nearby galaxies all that is necessary is to patrol a few hundred and you will bag a supernova every 2 months. Dream on! Unfortunately, supernova patrolling is highly competitive, and there are many other professional and amateur patrols out there. In practice, the dedicated amateur patrollers find that they average one supernova for every 4000 galaxy images exposed, and that is with a very intensive imaging and checking regime. By intensive we mean that if a night is clear, hundreds of galaxies are patrolled and the effort is repeated on subsequent clear nights. This strategy is essential to beat the other patrollers. Just patrolling a few dozen galaxies every clear night for a year might net thousands of galaxy images, but the imaging rate will just be too slow to snatch a discovery from the other patrollers who, combined, are scouring 10,000+ galaxies every week!

Of course, as with any competitive 'sport' there are a few tricks that can give you an advantage, apart from hiring a hitman to hobble the opponents. Professional automated patrols tend to avoid imaging when the sky is brightened by the Moon, especially when it is within a few days of full. This is primarily because much of the initial checking carried out by professional systems is automated; a master image is subtracted from a patrol image, and anything left over flags a potential supernova that can be investigated in more detail.

If the sky is bright every image will look different from the corresponding master. So, rather than avoid the Moon, some amateur patrollers take maximum advantage of it, patrolling even on the night of the full Moon and targeting the galaxies that its glow whited out the night before. Bright galaxies can easily be patrolled using this method, as even an exposure of a few seconds, with a CCD camera on a 300-mm telescope, can reach stellar magnitudes of 15 or fainter. Also, many supernovae occur close to the central bulges of galaxies and can easily be overlooked at first glance, especially if the imaging software applies an automatic contrast stretch to the image, which may appear to have burned out the galaxy core. Using a short exposure of a few seconds it is fairly rare for even a bright galaxy core to be saturated. Lowering the image contrast or disabling the auto-stretch will show the full dynamic range of the original and reveal any supernovae that might otherwise have been missed.

Just because you have a telescope that can slew across the sky does not mean you should spend the whole night with the telescope moving huge angular distances between the brightest Messier or Caldwell catalog galaxies. Slewing takes up vital patrolling time when you are trying to farm as many galaxies as possible. It also takes its toll on your worm and wheel, even on the best systems, if hundreds of galaxies are being patrolled night after night. In areas such as Virgo, Coma Berenices, or Ursa Major, where bright galaxies abound, you may only have to slew a degree or two to bag the next target. Indeed, if your telescope field is wide enough, you may get more than one galaxy in the telescope field. In 2000 Miles Paul of the Webb Society compiled an 'Atlas of Galaxy Trios' in which, using Digital Sky Survey (DSS) images, he displayed trios of galaxies down to magnitude 16.5 where no two galaxies were more than 10 arcminutes apart. The atlas is a useful resource for the supernova patroller who wants to minimize slewing time when using a narrow field of view.

Obviously patrollers need extensive galaxy lists so they are well prepared when the next clear night arrives. With a German Equatorial Mount (GEM), like the Paramount ME (the favorite choice of many top patrollers), crossing the north–south meridian line needs to be avoided, if possible. If the telescope is slewed so that the optical tube would end up lower than the counterweights, and might hit the plinth or pier, the control software will execute a 'meridian flip' to 'normalize' the system and avoid a tube/plinth collision. In other words, the polar and Dec. axes do a 180-degree rotation (arranged so the optical tube never points down during the slew). All very clever, but it wastes valuable patrolling time, especially if it happens more than once.

Patrolling close to the meridian and the zenith is a sensible plan, as the limiting stellar magnitude will be better at high altitudes, and light pollution/haze is minimized too. However, near, but not crossing, the meridian is the best plan. Some patrollers prefer to move eastward in increasing right ascension as the night moves on. For example, at an average imaging rate of 30 galaxies per hour, you could cover 30 galaxies as you also move east by 1 hour, slewing 2 minutes of R.A. (half a degree on the celestial equator), on average, between galaxies. This strategy would mean the telescope varies little in physical position throughout the patrol – a big advantage if you are observing through a dome slit where the dome is not automated.

The British patroller Mark Armstrong prefers searching via constellation-based lists, thus he sweeps in R.A. and Dec. throughout the night. His countryman Tom Boles prefers to sweep along lines of declination up and down the meridian. Simply searching as many big face-on galaxies, that are not too distant from our own Milky Way Galaxy, as possible, is the best approach for the beginner. You need to cover as many stars as possible, which means as many galaxies as possible, with a huge number of stars in each.

How far away should your galaxies be? Well, if you can get to magnitude 19, you can confidently discover magnitude 18 supernovae. For the faintest core-collapse supernovae, with absolute magnitudes around −17 this translates into a distance of 100 megaparsecs or 326 million light-years. Using the currently most popular Hubble recession velocity of 70 km per second per megaparsec it also translates into a galaxy recession velocity of 7000 km per second, a parameter that is sometimes quoted instead of the galaxy's distance. So you could use 7000 km per second as your galaxy recession velocity threshold. An alternative, rule-of-thumb, way of looking at this is to say that a typical core-collapse supernova at maximum will be about 2 magnitudes fainter than the galaxy it appears in. So if you are happy that you can discover supernovae down to mag 18 you can include any galaxies down to magnitude 16 in your search. Of course, Type Ia's in those galaxies will be well above your limit, at least on clear and dark nights. Again, patrolling galaxies with lots of stars, that is, huge galaxies, will increase your chances considerably. For example, in our Local Group the face-on galaxy M33 and the famous (but not face-on) galaxy M31 (the Andromeda Galaxy) might seem like similar bets for a patroller with a wide field system. However, with maybe 10 times as many stars, M31 is by far the better bet, statistically at least.

The Competitors

Many of the supernova searches carried out by the relentless sky-scouring machines of professional patrols are farming faint supernovae in huge numbers. They simply image clusters of faint, distant galaxies with large format CCDs, collecting multi-gigabytes (and even terabytes) of data and then powerful software scours the images, assisted by human checkers. It might seem that such patrols

Figure 4.9. British supernova hunter John Fletcher poses with the world's most productive supernova discovery machine, the Lick Observatory's Katzmann Automated Imaging Telescope. It may look like a wigwam, but by the end of 2007 it had discovered more than 650 supernovae in 9 years, peaking at 95 in 2003 alone. Image by kind permission of John Fletcher.

Figure 4.11. Tom Boles, discoverer of 108 supernovae, with two of his three Celestron 14/ Paramount systems. Image by kind permission of Ron Arbour.

discoveries were credited as joint (i.e., near-simultaneous) discoveries, made within minutes of each other, despite thousands of galaxies being trawled. However, unlike the Puckett and Schwartz/Tenagra patrols, the British patrollers worked alone, under (mainly) cloudy skies.

Supernova patrolling can be a stressful business, and quite often you are thwarted from a really bright supernova by cloud or by just plain bad luck. On July 26, 2006, Tom Boles observatory was struck by lightning, and all three

Figure 4.12. Mark Armstrong, discoverer of 73 supernovae, in 2003 with all three of his Celestron 14/Paramount systems. Image by kind permission of Mark Armstrong.

Paramount/Celestron 14 patrol telescopes were damaged by the surge. He was not back in action for 4 months, and even then only in a limited capacity! In Arizona, the observatory of the well-known American comet imager Mike Holloway was struck by lightning only 9 days later, destroying all his equipment. Check that insurance small print; these things do happen!

However, not all the supernova patrollers are happy in each other's company or even enjoy the nightly ritual. They are driven to discover supernovae and the drive, in many cases, is competitive. After a run of clear skies all patrollers pray for a cloudy night, at least, all the ones we know! And there are rivalries. There is at least one case where a discoverer claimed he had his supernova stolen by another patroller. How? Well, discoverer number 1 phoned discoverer number 2, asking if he had a recent image of a galaxy in which he had a new suspect. Discoverer number 2 looked at his own image from the previous night, agreed there was a new supernova and then, as soon as the conversation ended, e-mailed the Central Bureau claiming he was the discoverer, with discoverer number 1 named as the co-discoverer: that is the version of the story according to discoverer number 1. According to discoverer number 2, the phone call of discoverer number 1 delayed him checking that very galaxy image himself, and the Central Bureau re-worded his own co-discovery claim so he mysteriously appeared as the principal discoverer. A short time later discover number 1 told me: "All's fair in love and war. . . and, it would appear, supernova patrolling!" In addition, a manic obsession with patrolling can play havoc on the observer's mind and his relationship with his family. More than one divorce may be due to supernova patrolling. You have been warned! Astronomy tends to be a far more relaxing hobby if you have your own little niche where you are not competing with others and where your best friends are interested in a slightly different aspect of the hobby. Sometimes observing supernova patrollers at war can be a hobby in itself!

Of course, no mention of amateur supernova hunting can omit the man who inspired all the rest, Bob Evans of New South Wales, Australia. This remarkable observer has discovered 46 supernovae, all of them visually, simply by memorizing the starfields around each galaxy, for around 1000 galaxy fields. Each galaxy is studied for a matter of seconds before starhopping to the next, a process that averages out at roughly one galaxy checked per minute. In many ways Evans' role of discovering supernovae from the southern hemisphere has been passed to Berto Monard, whose observatory is sited at the top of a 1590-m-high ridge 40 km east of Pretoria. At the time of writing Berto has discovered 58 supernovae.

It goes without saying that false alarms will not make you a popular person in supernova discovery circles. The correct procedures for reporting a supernova discovery are detailed on the CBAT website at http://www.cfa.harvard.edu/iau/Dis coveryInfo.html. CBAT also has a web page for checking minor planets (asteroids) in the vicinity of a galaxy that may be mistaken for a supernova; see: http://scully. harvard.edu/~cgi/CheckSN.com. In addition CBAT also has a 'faint unconfirmed supernova' page for mag 19 and fainter possible discoveries (i.e., professional supernova suspects) at http://www.cfa.harvard.edu/iau/CBAT_PSN.html

The golden rule in supernova searching is to build up a set of master galaxy images with the same telescope and CCD camera that you use for patrolling, but making sure the master images are as deep as you can possibly ever go with the patrol images. Variable stars and asteroids can easily fool you, as can 'hot pixels,' which occur on every CCD image. Essentially, two separate images of the supernova, taken on consecutive nights, will be the minimum that will be acceptable as proof, even from a previously successful amateur. From a newcomer to the field, a confirming image from an experienced supernova discoverer will usually be needed. An additional trap for the unwary is discovering a supernova that has already been found. Most bright supernovae are discovered within days of them exploding and will appear promptly on the CBAT 'recent supernovae' page at http://cfa-www.harvard.edu/iau/lists/RecentSupernovae.html.

Hardware for Patrollers

In recent years the most successful supernova patrollers have simply been those with the spare time to dedicate to patrolling and, in many cases, the money to buy the most reliable 'Go To' mountings. Maybe you have even heard the phrase 'a supernova costs $1000'! The top choice among these patrollers has invariably been the $12,500 (2007 price) Paramount ME, often carrying a Celestron 14 (0.35-m f/11 Schmidt–Cassegrain) payload, and a 512×512 CCD camera, such as SBIG's ST9XE; all told, a $20,000 investment. With some amateurs operating a trio of such systems things start looking prohibitively expensive!

However, amateurs equipped with far more modest systems have also been successful. In Britain Ron Arbour has discovered 18 supernovae with a single (if highly modified) 30-cm LX200 and a homemade 40-cm Newtonian, and Berto Monard in South Africa uses an identical 30-cm LX200. The reason that the C14, combined with a 20-micron pixel detector, works so well is due to both light grasp and resolution. In unguided 60-s exposures such a system can reach magnitude 19, and with a 3900-mm focal length (355-mm aperture at f/11) the corresponding image scale of 1 arcsecond per pixel can weed out those supernovae close to the nucleus of a galaxy. In addition, the 8.5 arcminute field of such a system is just large enough to contain 99 percent or more of all the galaxies you are likely to patrol. However, CCD detectors now have considerably more pixels than at the start of the amateur discovery bonanza (the late 1990s), so maybe a less expensive rethink is necessary. Even the most basic Digital Single Lens Reflex Camera (DSLRs) have 2000×3000 pixel arrays, and at a scale of, say, 2 arcseconds per pixel, a 1.1×1.7-degree field could be captured. With DSLR pixel sizes of 6 or 7 μm being typical, this translates into a system with a focal length of about 600 or 700 mm, so a 150-mm f/4 or f/5 Newtonian might perform well. In regions such as the Coma-Berenices/Virgo cluster, where bright galaxies seem to appear in almost every telescope field and supernovae will be 14th or 15th magnitude, such an affordable system could be attractive to many patrollers.

Software for Patrollers

Software for supernova patrollers can essentially be split into two types, and, frankly, there is not much available in either category. The first type of software that is required is something to simply automate the laborious task of slewing and imaging so many galaxies per hour. This could be done manually, and if you are simply interested in a casual, even relaxing patrol where galaxies are checked visually before slewing to the next one, it might work. However, if you are churning through galaxies, and especially if the telescope is left unattended, you will want imaging and slewing to be integrated and automated. The most popular package for this (by far) is Software Bisque's Orchestrate scripting system, which allows that company's CCDSoft imaging package and The Sky planetarium software to work together. You simply open both applications and set the pre-prepared script file running, with commands like

```
SlewToObject NGC4438
WaitFor 5
ImageThenSlewTo 60.0 NGC4425
WaitFor 5
ImageThenSlewTo 60.0 NGC4388
WaitFor 5
ImageThenSlewTo 60.0 NGC4406
```

This sample script is fairly self-explanatory. It consists of three different command types: a slewing command, an image and then slew command, and a wait command. The wait command is set for 5 s so that the telescope has settled down after a fast slew to a target. Even with a Paramount ME, decelerating from a high-speed slew to stable tracking does not take place instantaneously! With the Paramount ME being the most successful supernova patrol telescope among serious patrollers, the fact that its manufacturer, Software Bisque, also wrote The Sky, CCDSoft, and Orchestrate, is a powerful reason for using this software suite for patrolling. Frankly, nothing else comes close.

Checking Software

This is the real bugbear for the patroller. You have to check an average of 4000 galaxy images per discovery. Sometimes you may just check 100 and get one quickly. Other times you have to check 20,000, believe you will never get another, and lose the will to live (so the top patrollers say). These are not trivial statistics. Most people will have had enough after the first 50 galaxy images. If only there was an automated supernova searching system?

Well, maybe not, as then everyone would be discovering supernovae. The professionals use software to flag up likely supernovae simply by subtracting the master reference image from the patrol image. Then the images are scoured by teams of students or researchers. Although the Puckett and Schwartz patrols do have plenty of amateur collaborators and checkers, in the US and worldwide most amateurs just check their images alone, a demoralizing mountain of work with no guarantee of success at the end. It is this daunting prospect that kills off most discoverer's aspirations.

To alleviate the pain, what is really required is a slick and intuitive software-checking system that loads the master and most recent patrol image for each galaxy, aligns them to pixel accuracy, and blinks them in rapid succession at high and low contrast (the low-contrast image will weed out any suspects close in to the galaxy nucleus). Sadly, to this author's knowledge, there is no supernova checking software that really delivers the goods, one that throws every patrol image from the last night's work in your face and flawlessly blinks it with the master, then moves to the next image. There is a blink comparator in CCDSoft, but the loading and comparison process is manual and not slick enough for easily blinking hundreds of images one after the other. The same argument applies to the blinker in Richard

Berry's AIP4Win. There are amateurs who use Microsoft Powerpoint to blink images, as if they were consecutive slides in a slide show. However, this is a tedious method, too, and, of course, Powerpoint only understands JPEGs, BMPs, and TIFs, not astronomical Flexible Image Transport System (FITS) format images.

Ajai Sehgal, who is a member of Tim Pucket's Supernova Search Team, has written a useful supernova search tool that is now incorporated into Diffraction Limited's Maxim DL image processing software. Sehgal's software requires you to have 'before and after' images with the same file names in separate 'before and after' directories. Selecting auto-blink will align and blink images of the same image scale. Pressing Next (or Back) will move to the next pair in the two directories that need checking. Perhaps Sehgal's supernova search tool is the best option, at least if you have Maxim DL (which can control the telescope and CCD camera, too). Maxim DL does cost about $400, though. Alternatively, and free (!), you could download Dominic Ford's software Grepnova, which can be found at his Cambridge University website at www-jcsu.jesus.cam.ac.uk/~dcf21/software.shtml. Grepnova aligns patrol and master images and blinks them and has been successfully trialed and used by Britain's supernova patroller Tom Boles to discover the supernovae 2005dj, 2005ej, 2005io, 2005ip, 2005lx, 2006A, 2006ao, 2006ap, 2006aq, 2006ar, 2006bk, 2006bl, 2006bo, 2006cr, 2006ow, and 2006ss.

Astrometric Software

Whether you are a nova patroller or a supernova patroller, the chance is that at some point you will need to carry out some astrometry – measure the precise position of an object you think you may have discovered, with reference to nearby stars. The best software for carrying out astrometry is called Astrometrica, written by Herbert Raab. The software is free for a trial period (shareware), and very affordable if you purchase it. More importantly, it is intuitive to use and very accurate. Ideally your CCD image needs to be in the astronomical 'FITS' format, which almost all astronomical CCD cameras support, but some other image formats used by SBIG cameras are supported, too.

With any astrometric software you need access to a detailed digitized star catalog, as it is only by reference to such a catalog that an object's precise position can be measured. The UCAC2 catalog, produced by the U.S. Naval Observatory, is the most useful in this regard with the entire 2-gigabyte database, in three chunks, easily fitting onto a hard disk. A specific request to the U.S. Naval Observatory is needed to acquire it as the original 1000 CD production run soon dried up! However, the UCAC3, which fills in the missing stars above +45 Dec., is due out soon. Software Bisque's The Sky 6 contains the UCAC2 catalog and its 48 million stars and, used with CCDSoft, can carry out slick astrometry. But neither of these packages is cheap. The UCAC2 and the massive USNO B1.0 data are available for the downloading of specific sky areas at http://vizier.u-strasbg.fr/viz-bin/VizieR/

If you have chosen to use Astrometrica, you first click on File/Settings/Environment/Star Catalog and point the software to your preferred star catalog. You then load your FITS/SBIG format image into Astrometrica and, to get North at the top, you can flip the image vertically and horizontally if it has south at the top or is mirror-imaged. You then select Astrometry/Data Reduction and Click 'OK.' Astrometrica will auto-identify the field (using the FITS header info regarding approximate position and telescope scale) and WHAM; it identifies every catalog star in the field, instantly! Mouse-clicking on the object of interest then reveals the precise R.A. and Dec. (and the magnitude) of the object you are measuring. The Astrometrica home page can be found at www.astrometrica.at

Chapter 5

Active Galaxies

How can an object the size of a galaxy vary its light output in a period of weeks, or even days, to such an extent that amateur astronomers can easily record the variations visually (or with CCDs)? As the engineer Scotty aboard the starship *Enterprise* might have said to Captain Kirk, 'You canny break the laws of physics Cap'n.'

Without invoking superluminal science fiction velocities, if an object varies its brightness in a period of days, then the source of the variation must be light-days or smaller in diameter. In addition, where galaxy-sized objects hundreds of thousands of light-years across are concerned, the source of the variation needs to be intensely bright with respect to the rest of the object.

Our solar system, out to the orbit of Neptune, is 8 light-hours in diameter, which gives an idea of the sort of scale we are talking about. At the centers of these so-called 'active galaxies' (active galactic nuclei are usually abbreviated to AGNs) lie massive black holes that may have swallowed a billion solar masses and continue to swallow another solar mass every few weeks. Thus, the term 'active galaxy' is a bit misleading. The whole galaxy is not varying in brightness, only the center. Remember, a typical supernova is a star less than a light-minute in diameter, but when it goes off it can outshine its host galaxy. However, unlike a supernova, an AGN is constantly devouring material to produce a steady, if fluctuating, output across the electromagnetic spectrum (see Figures 5.1 and 5.2).

Broadly speaking, astronomers classify active galaxies into two types, namely starburst galaxies and the aforementioned AGNs. Amateur astronomers often image starburst galaxies but they observe the variations in AGNs. This chapter is primarily about AGNs. But, in case you are curious, I had better explain the term 'starburst'. Galaxies awarded this classification are, not surprisingly, ones that are experiencing intense star formation way beyond the average seen in a typical galaxy. In addition the star formation tends to take place in the center and/or in a relatively small area of the galaxy. Intense star formation galaxies can have areas of space only the same radius as the distance our Sun is from the bright star Altair (17 light-years) but with 100-million-Sun power being generated in each region by massive short-lived stars. AGNs frequently contain starburst regions, too, but their defining feature is an excessively violent output from a central supermassive black hole.

The power output of AGNs is staggering, typically 10^{39} W, and even 10^{41} W in the most extreme cases. Instinctively, you might imagine that black holes could not emit radiation; they are black, after all. However, the radiation emitted is as a result of infalling matter, well outside the region where it might pass through the nonreturn valve of the event horizon.

Although swallowing a dozen or more solar masses a year is a massive enough concept to grapple with, some astronomers think that a few of the largest quasars

M. Mobberley, *Cataclysmic Cosmic Events and How to Observe Them*,
DOI: 10.1007/978-0-387-79946-9_5, © Springer Science+Business Media, LLC 2009

Active Galaxies

Figure 5.1. The active galaxy NGC 4261 imaged by ground-based optical and radio telescopes as well as a high-resolution image by the Hubble Space Telescope (HST). The region within several hundred light-years of the central black hole is shown in the right-hand image. The entire galaxy and its radio jets, spanning over 100,000 light-years, is shown on the left-hand side. Credit: NASA/ STScI.

Figure 5.2. About 50 million light-years away in the center of the relatively nearby peculiar galaxy NGC 4438 (in the Virgo Cluster), a supermassive black hole causes a huge bubble of hot gas to emerge from the galaxy's core. Imaged by the Hubble Space Telescope. Image credit: NASA/ESA/ Yale University.

contain black holes that can potentially swallow 1000 solar masses per year, provided they are surrounded by enough matter to swallow. This latter point is crucial. Black holes and the environment around them do not generate energy of their own accord. They need fuel to feed on. Without fuel there can be no activity; a supermassive black hole on its own just distorts spacetime but does not show any violent activity. Astronomers think there is a black hole at the center of our own galaxy of around 3 million solar masses. They also think the Andromeda Galaxy, M31, may have a central black hole of some 150 million solar masses. However, in both cases there does not appear to be enough fuel feeding the black holes to classify the centers of these galaxies as AGN. In addition, even a 150-million-solar-mass black hole can look rather tame in the AGN world.

Super massive Black Holes

Current AGN theories require that supermassive black holes really do exist, although proving this to everyone's satisfaction is incredibly difficult outside our own galaxy. Even though these gravitational monsters will be huge, their diameters are well below the resolution of any instrument, at least when looking at galaxies tens or hundreds of megaparsecs away. Outside our own Milky Way and Andromeda, only the central black hole in M106 (35 million solar masses) has a weight of observational evidence to support its existence. Nevertheless, most astronomers are convinced that enormous black holes live at the centers of most galaxies, and when provided with enough 'fuel' can explain the behavior seen by professionals and amateurs, such as staggering variations in light and energy output on small timescales. When we are talking about the centers of active galaxies, black holes of several hundred million, or even several billion, solar masses are being considered. These objects can swallow a solar mass every few weeks and over the course of a year emit more than 10^{47} J of energy.

When an object falls into a black hole and passes beyond its event horizon it cannot return to this universe. To do so it would have to overcome an escape velocity greater than that of the speed of light. Thus, matter being swallowed by a black hole does not release a gigantic interstellar belch, whereby radiation is spewed out into the cosmos. We are not talking about superluminal indigestion here.

Admittedly, some readers may have heard of 'Hawking radiation' often alongside the rather glib phrase 'black holes aren't black.' This quantum peculiarity allows particle and antiparticle pairs to be created at the event horizon of a black hole, and for one particle to fall in while the other escapes, thus liberating radiation from the black hole. However, this is not what we are talking about here. AGNs generate energy by the infalling matter before the matter has fallen into the black hole. If an object falls from a distance of light-weeks straight into a supermassive black hole, it can enter the event horizon as it approaches the speed of light. This speed has been generated by a steady gravitational attraction as the object is pulled in over the vast distances of space.

All high school physics students will know that kinetic energy is equal to $\frac{1}{2}mv^2$ and, from Einstein, that the colossal nuclear energy released by turning the rest mass of an object into energy is mc^2, where c is the speed of light. If an object is traveling close to the speed of light and then collides with something it will therefore liberate colossal amounts of energy, a significant fraction of what converting it to pure energy would yield. But, in practice, supermassive black holes will tend to generate an

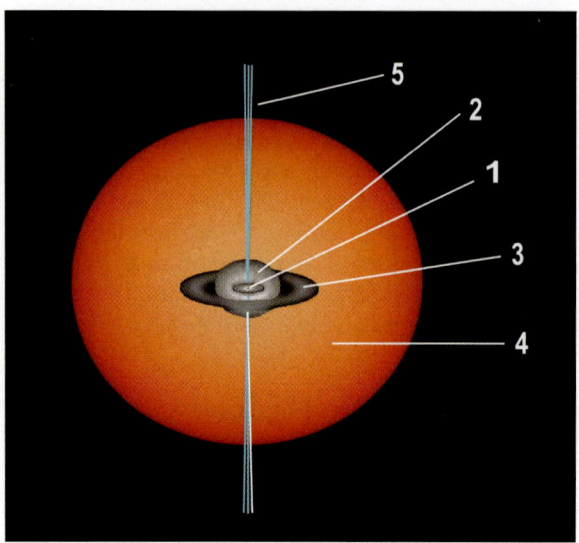

Figure 5.3. A block diagram of the central region of a typical active galaxy. Not to scale.

1. Supermassive black hole surrounded by an accretion disk with a radius of light-months.
2. Region roughly 5 light-years across, containing X-ray and infrared synchrotron radiation, and also containing broad-line-emitting gas clouds, maybe 10 light-minutes across.
3. Warm 'doughnut-shaped' region of highly absorbent molecular dust and gas, several hundred light-years in diameter.
4. Region containing narrow-line emission gas clouds, maybe tens of thousands of light-years in diameter.
5. High-speed, or relativistic, jets seen in radio-noisy active galaxies. These often end in a huge radio-emitting lobe.

accretion disk, not dissimilar to that surrounding a cataclysmic variable star, but on a stupendously larger scale, with a radius of light-weeks or months. The general structure of the core region of an AGN is shown in Figure 5.3.

Material in the inner accretion disk will, via collisions and liberation of kinetic energy, reach a million degrees K or more, and such high temperatures can create colossal amounts of energy in such a chaotic environment. A hot and rarefied region envelops the disk, and X-ray radiation plus infrared synchrotron radiation is emitted. Jets channeled by powerful magnetic fields can form within the plasma of this region and spray out along the rotation axis of the system at high speeds. Where these jets attain a significant fraction of the speed of light (i.e., they are moving at relativistic speeds), a strong radio emission will be observed from the AGN. Astronomers think that these examples may be associated with a very fast rotating black hole. Of course, the amount of fuel available will also have a significant bearing on the violence of the AGN.

Active galaxies come in various types. In simplistic terms, they are often summarized thus: quasars are ultradistant and ultrapowerful; blazars are closer and with a jet of radiation pointing straight toward us; and Seyferts are typically barred or tightly wound spiral galaxies with a small, very bright nucleus and come in two distinct types.

But, perhaps surprisingly, the main difference between different AGNs is simply the angle that we view the object from (see Figure 5.4). For nonrelativistic jet/radio-quiet AGNs we see a quasi-stellar object (QSO) or Type 1 Seyfert if looking

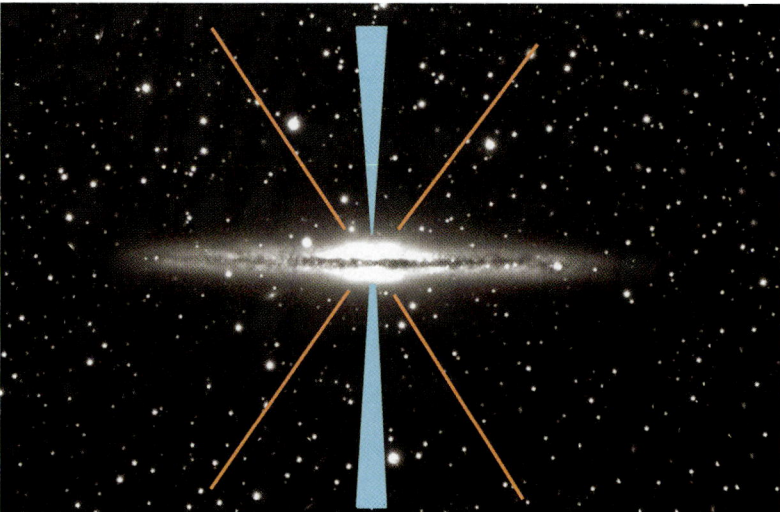

Figure 5.4. What a distant observer sees when looking at an active galaxy largely depends on the viewing angle. For AGNs with jets an observer looking straight down the narrow central cone (solid in the diagram) will see a blazar. Further out, as far as the single lines in the diagram, a quasar will be seen. Further out still, with the observer looking more along the plane of the AGN, a radio galaxy will be observed. For AGNs that have no jet and are radio-quiet the single lines in the above diagram will roughly represent the point at which a Seyfert Type 1/Quasi Stellar Object is seen (galaxy more face-on than edge-on) or a Seyfert Type 2 is seen (galaxy more edge-on than face-on).

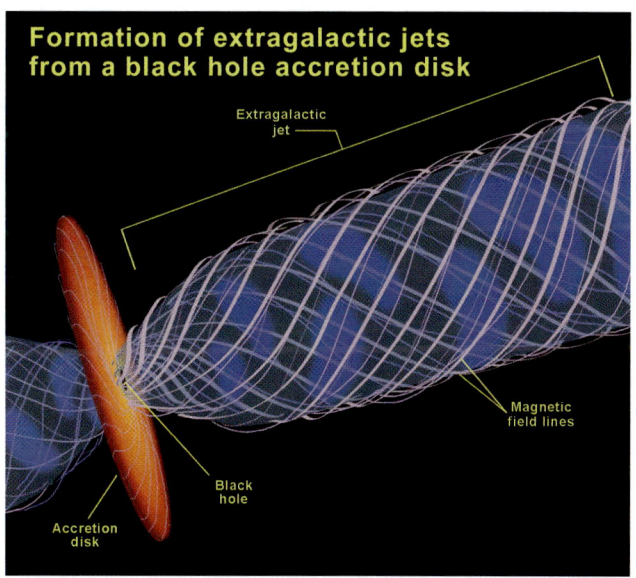

Figure 5.5. Jets emitted by radio-noisy active galaxies are thought to be made possible by powerful magnetic fields channeling an outpouring of subatomic particles from the black hole into a magnetically constricted jet. This particular artist's impression was drawn to represent the jet in the galaxy M87. Credit: NASA/ESA and Ann Field (Space Telescope Science Institute).

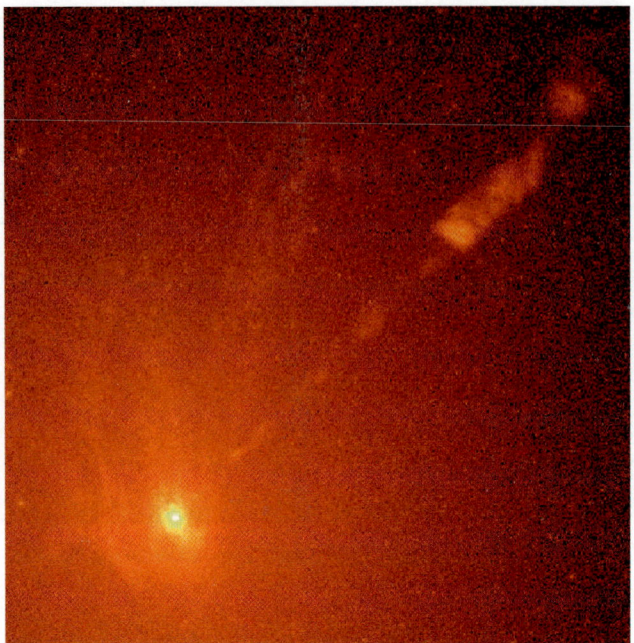

Figure 5.6. The jet from the well-known galaxy M87, imaged by the Hubble Space Telescope. Credit: STScI/NASA/ESA.

along the rotation axis or up to 45 degrees off-axis. But if we are looking edge-on to the disk, up to 45 degrees away from the disk, we see a Type 2 Seyfert. For relativistic jet/radio noisy AGNs we see a blazar if we are looking right down the jet (lucky us!); further out, say 30 degrees, we see a quasar. Further out still we might see a broad-line radio galaxy, and then as we approach within 45 degrees of viewing the disk plane edge-on we see a narrow-line radio galaxy. Jets are an optional extra with active galaxies; they are thought to be created when intense magnetic fields channel an outpouring of subatomic particles from the black hole by magnetic constriction (see Figure 5.5).

The jet in the relatively nearby Messier Galaxy, M87 (see the Hubble image in Figure 5.6), makes a great target for a dedicated deep-sky imager. Even with modest CCD equipment it can be captured with exposures of 60s. By stacking multiple images and keeping as much of the inner galaxy core as possible below saturation, much detail can be resolved in the jet. M87 is a radio galaxy (radio designation Virgo A) only 60-million-light-years distant. Professional astronomers have determined that the jet traces back to a region less than light-weeks in diameter at the core of this relatively featureless galaxy. The jet spans a distance of 6000 light-years.

Confusing Classifications

It would be nice in astronomy, or in any field, if you could simply classify every object in a box. However, there are starburst galaxies, low-luminosity AGNs, high-luminosity AGNs, and a host of variations. But, generally speaking, AGNs feature rapid variations in brightness compared to the centers of 'normal' galaxies, strong optical emission lines, and, in some cases, jets of material seen ejected from the nucleus. Giant radio lobes, hundreds of thousands of light-years from the AGN, are often seen too. In addition, these galaxies frequently have absurdly high outputs in the infrared, ultraviolet, and X-ray regions.

AGNs are not the brightest objects visible in amateur telescopes; however, a skilled visual observer can easily monitor a dozen candidates with a 30-cm instrument, and with CCD equipment much smaller telescopes can carry out useful monitoring work. In many cases AGNs are studied at radio or even X-ray/gamma-ray wavelengths by the professionals, but not in the optical spectrum; therefore, observing AGN is a really valuable pursuit, as the amateur can fill in the wavelength gap and even alert professionals to a rise in brightness. The Veron catalog is the definitive list of active galaxies discovered by professional astronomers, and the latest version at www.obs-hp.fr/www/catalogues/veron2_10/veron2_10.html contains a staggering 23,760 quasars, 608 BL Lac objects, and 5751 active galaxies (of which 2765 are of Type Seyfert 1). Needless to say, only the brightest examples are within the range of amateur astronomers.

Strictly speaking, quasars and QSOs are subtly different: quasars have strong radio emissions, and QSOs do not. Despite this, most amateur astronomy literature does not distinguish between the two, probably because few amateurs are interested in radio emissions, unless they have a huge dish in their backyard. There are other categories too, such as LINERs (low-ionization nuclear emission regions), NLXGs (narrow-line X-ray galaxies), BALQSOs (broad absorption line quasi-stellar objects), RQQs (radio-quiet quasars), FRs (Fanaroff–Riley radio-noisy objects), NRLGs (narrow-line radio galaxies), ULIRGs (ultraluminous infrared galaxies), OVVs (optically violently variable quasars), HPQs (high-polarization quasars), BLRGs (broad-line radio galaxies), LPQs (low-polarization quasars), and RLQs (radio-loud quasars)!. But this book is really aimed at amateur astronomers who will want to go out and observe visually or with CCDs at an image scale of 1 or 2 arcseconds per pixel.

We will now look at a few specific examples of objects worthy of amateur scrutiny, subdivided into the categories in which they are normally classified. Charts for finding and estimating the magnitudes of these popular AGNs can be found on the AAVSO site at www.aavso.org/observing/charts/ or the AAVSO variable star plotter site at www.aavso.org/observing/charts/vsp/.

Quasars

"Twinkle, twinkle quasi-star

Biggest puzzle from afar

How unlike the other ones

Brighter than a billion suns

Twinkle, twinkle, quasi-star

How I wonder what you are."

—George Gamow, "Quasar," 1964

Quasars are often considered as the most powerful category of AGN and are all a long way from our galaxy. The closest example is around 800 million light-years away and the farthest, at the edge of the visible universe, roughly 13 billion light-years away. Objects this far away are often allocated a redshift, or 'z,' value rather than a distance. However, a light-travel time to these objects can be calculated from z as long as some assumptions are made.

Of course, the light-travel time implies the distance in light-years that the light has traveled to get to you, but not the distance that the object would now be, which would be much greater. Table 5.1 shows the light-travel time for various values of z.

At the present time the farthest objects imaged have redshifts of about 6.5, so we are seeing objects only a billion years after the Big Bang. In practice, the situation is even more confusing than this, because the universe appears not to have been

Table 5.1. Light-travel time (giga years or billions of years) for various values of z

Redshift z	Light-travel time (Gyr)
0.02	0.272
0.04	0.536
0.06	0.793
0.08	1.043
0.1	1.286
0.2	2.408
0.4	4.256
0.6	5.694
0.8	6.826
1.0	7.731
1.2	8.462
1.4	9.061
1.6	9.557
1.8	9.972
2.0	10.324
3.0	11.476
4.0	12.094
5.0	12.469
6.0	12.716
7.0	12.888
8.0	13.014
9.0	13.110
10.0	13.184
1100*	13.665

*The final 'z' of 1100 is the redshift of the cosmic background radiation itself – in effect, the Big Bang at 13.7 billion years (and light-years) distant. A 'z' of this size indicates how far the visible radiation of the Big Bang has been shifted in wavelength, that is, into the microwave region with a temperature of 2.7 K. For the table we have assumed a Hubble constant, H_o, of 71 km per second per megaparsec, $Omega_M=0.27$, and $Omega_\lambda=0.73$; these are parameters governing the expansion rate, relative mass density, and cosmological constant density that were favored in 2008

expanding at a steady rate throughout cosmological time. Nevertheless, Table 5.1 does give a reasonable view of how 'z' and light-travel time are thought to be related by the vast majority of astronomers.

3C 273

Despite the extremely dull name (meaning this object is the 273rd listed in the 3rd Cambridge catalog of radio sources), 3C 273 (see Figure 5.7) is an extraordinary object and has a place in the history books too. 3C 273 was the first object to be called a QSO, or quasar, an unbelievably luminous object at a distance of billions of light-years. It was not the first quasar to be discovered (the quasar 3C 48 was already known to be an enigma), and neither was it the 273rd, as that number is simply a result of its right ascension, or celestial longitude. However, it was the first quasar radio source to be recognized for what it really was. The quasar 3C 48 was actually the first to be cataloged, but 3C 273 was the first to be optically identified and then have a spectra taken.

Because radio telescopes using single dishes do not have especially high resolution, a crude but ingenious method was used to work out what visible light object corresponded to 3C 273. The Moon was used to occult the radio source! By timing

Figure 5.7. The Quasar 3C 273 and its jet, imaged by the Hubble Space Telescope. Credit: STScI/ NASA.

precisely when the radio signal disappeared and reappeared as the Moon occulted 3C 273, an accurate fix on its location could be established. The source was found to coincide with a stellar source, and when astronomer Marteen Schmidt at the Mount Palomar Observatory measured the spectra of this modest looking object, it turned out to have a redshift of 0.158, placing it at the colossal distance of 2 billion light-years, with a velocity away from us of roughly 15 percent of the speed of light. For an object as bright as 12th magnitude (12.3–13.3 is the usual range), this was astonishing.

To put this into context, the absolute magnitude (i.e., the magnitude at a distance of 10 parsecs or 32.6 light-years) of 3C 273 had to be brighter than the observed magnitude of 3C 273 by a factor of the square of (2 billion/32.6), that is, 3.8×10^{15}. In magnitude terms this works out to be 39 magnitudes! So, taking the most common brightness of 3C 273 as about 12.9, if it was only 10 parsecs away it would be brighter than magnitude –26! This is virtually the same brightness as our Sun appears to us, only 8 light-minutes away. Put another way, 3C 273 is as luminous as 2 trillion Suns (2,000,000,000,000). If this were not staggering enough, the luminosity is variable across all wavelengths (visible, infrared, ultraviolet, radio, gamma, and X-ray) and can be detected even across consecutive nights with amateur equipment. On rare occasions it has brightened to magnitude 11.7. This makes 3C 273 the brightest quasar and the most distant object that can be seen visually through a modest amateur telescope.

3C 273 is certainly not the most luminous quasar, either. The objects known as APM 08279+5255 and HS 1946+7658 are thought to have absolute magnitudes close to –30, allowing for gravitational lensing. Thus, these objects would shine as brightly as our Sun, 8 light-minutes away, even if they were removed to a distance of about 140 light-years. Mind boggling!

PKS 1749+096

Sometimes also known by its Ohio designation of OT+081, PKS 1749+096 is an interesting object for visual and CCD-equipped amateurs to study. It lies in the constellation of Ophiuchus, roughly midway between the bright, 2nd-magnitude-star Rasalhague and the 9th-magnitude-planetary-nebula NGC 6572. With a redshift of z=0.322, its light has been traveling for over 3 billion light-years to get to us, and it can boast a typical absolute magnitude of –24. PKS 1749+096 is usually classed as a quasar, although a BL Lac classification is sometimes used. To amateur observers it is, essentially, stellar, although the galaxy it resides in does have a catalog name of LEDA 84879. At first glance this object did not look like a worthy quasar for amateurs to study. Even now its SIMBAD entry gives it a B magnitude of 17.8, and various catalogs list it as varying between about 15.8 and 18.4, in other words, purely a CCD target. However, in 2007 the quasar went ape and was detected as bright as magnitude 14.1 by Stefan Karge at the Taunus Observatory, Frankfurt, in April of that year. For the rest of that object's observing season (until August) it only dipped (briefly) below magnitude 15 in mid to late July. This behavior is unusual for quasars or any form of AGN. Normally, outbursts are rather brief, but this one was significant and endured for months, at five times above its normal brightness level.

Blazars, OVVs, and BL Lac Objects

The name BL Lac originates, like many names in variable-star nomenclature, from the defining object of the class. A variable-star designation is usually made up of two letters (or the letter V and a number) followed by the constellation abbreviation. In the case of BL Lac objects the defining object is, therefore, BL Lacertae, an object within easy range of amateur instruments. The more evocative term 'blazar' encompasses both BL Lac objects and OVV quasars. Needless to say, it is not quite as clearcut as this; as we have seen, trying to box objects into specific categories rarely works in astronomy. There are intermediate objects that have a mixture of OVV and BL Lac characteristics, and the behavior of a few is confused by gravitational lensing effects.

Blazars are different from the superdistant common or garden variety quasars in that they have a jet that just happens to be pointing toward our line-of-sight/galaxy (and, quite possibly, another jet pointing in the opposite direction). Thus, we see rapid variations in a galaxy's radiation output because we are looking right into the active heart of the object. In general, the distinction between OVVs and BL Lacs is that the OVV quasars are ultrapowerful radioemitters, whereas the BL Lac objects are not, even though they do emit radiation right across the spectrum from radio to X-ray frequencies. However, the term OVV is rarely used, and quite often the terms blazar and BL Lac are used to mean the same thing.

The spectra of BL Lac objects often show no emission features, except when the object is faint, hinting that those features are simply swamped when the object is bright. Once again the powerhouse is material falling into a massive black hole and the accretion disk surrounding this object. Astronomers think this active region is probably only about a light-day across, in a galaxy that may span hundreds of millions of light-years in diameter. A region of opaque, hot gas many light-days across may surround the region. A useful blazar web page can be found at http://astro.fisica.unipg.it/blazarsintheweb.htm.

Both 3C 279 and BL Lac itself are probably the two most widely observed blazars, both by amateurs and professionals. 3C 279 is in the OVV category, and was the first active galaxy in which material, apparently traveling faster than light, was observed! This would be straight out of a science fiction novel if the material really was traveling at superluminal velocities.

However (and sadly, perhaps), astronomers soon came up with a good explanation that avoided faster-than-light travel, although a rather different one from the explanation for GK Per's apparently superluminal nova shell that we saw in Chapter 2. The jet emitted from an active galaxy's nucleus consists of highly luminous blobs of material that are visible in radio telescope images (using several radio telescopes to increase the baseline and resolution is standard practice), and the blobs can be seen to move over months and years despite the enormous distance to the object. These highly luminous blobs can travel at significant fractions of the speed of light, but not more than the speed of light. However, measuring the actual speed of one of these blobs is fraught with complications, even disregarding the complexity of using multiple radio telescopes linked together. If two measurements of a blob's position are made, say, one year apart, the blob will be significantly closer to Earth when it is measured for the second time. In fact, more than one year would have elapsed at the blob between the first and second measurement points, as the blob was further back in the cosmological past at the time of the first measurement.

A blob can travel further than you anticipated if it has more time to travel in, and when relativistic effects are included on a jet that is pointed very close to our line-of-sight, the apparent speed of the jet can appear to exceed the speed of light by many times. Sadly, amateurs cannot study these effects, but they can certainly observe blazar 3C 279. Let us have a closer look now at 3C 279 and BL Lac from the amateur's perspective.

3C 279

Like 3C 273, 3C 279 is also found in the constellation of Virgo, just 11 degrees southeast of its brighter counterpart. It lies roughly 5 degrees SSE of the famous double star Gamma Virginis, also known as Porrima or Arich. Like 3C 273, 3C 279 lies at a staggering distance from us and has an awesome luminosity; its absolute magnitude can exceed 27 at its peak. It varies in brightness between 13th and 16th magnitude. Over a long timescale of a couple of decades this AGN tends to spend periods ranging from 6 months to several years in a bright state, and when in this bright state brief flares in brightness, lasting a week or several weeks, are often observed. During these very active periods professional astronomers, and amateurs with a photometric capability, have unequivocally detected nightly variations as large as 0.5 magnitudes in 24 hours, or 0.2 magnitudes in 5 hours.

BL Lac

BL Lac, after which an entire group of active galaxies is named, lies in the northern constellation of Lacerta, the lizard (22 h 2m 42.8s +42° 16′ 37″), and thus is available all year round for far northern hemisphere observers; in February, it is poorly placed and low down (near the north horizon) at midnight. It sits roughly midway between Deneb in Cygnus and Alpheratz, the top left star in the Square of Pegasus. In late 1999 BL Lac displayed its brightest outburst for 30 years, peaking at just brighter than magnitude 13. Only two years earlier it had experienced a significant outburst at optical and X-ray wavelengths. However, for much of its life it is only around magnitude 15, and magnitude 16 is possible at the lowest points. At these times it is below the visual range of most amateurs, even those with 40-cm instruments.

Markarian 421

The active galaxy Markarian 421 (see Figure 5.8) is usually regarded as the nearest blazar (roughly 430 million light-years away). It is also known by its galaxy designation of UGC 6132. It lies close to the southwestern Ursa Major border with Leo Minor, at 11 h 4.5m,+38° 13′. A magnitude 6 star (GSC 30102500) lies just 2 arcminutes to its north-northeast, and the half-degree-wide field is very distinctive, with the bright star forming the tip (with a fainter star) of a diamond shape. Unfortunately, the nearby magnitude 6 star dazzles the eye, but CCD images, of course, are not prone to this problem. However, there is a shortage of good photometric reference stars in the immediate vicinity, so accurate magnitude estimates can be difficult to obtain.

Figure 5.8. The field of Markarian 421, an AGN monitored regularly by amateur astronomers. The image is 15 × 15 arcminutes wide with north up. Markarian 421 is the small, but bright fuzzy object in dead center. It has a smaller elliptical galaxy to its top left. The two bright stars in the image are of magnitudes 6.0 and 7.5. Credit: STScI/DSS.

Markarian 421 has experienced two significant outbursts, namely, in 1982 and 1992. On both occasions it reached 12th magnitude, and the latter outburst was prolonged with a relatively slow fade. However, the blazar has experienced much quieter and rather boring phases, such as from 1983 to 1988, after which it exhibited 3 years of relatively short-duration outbursts. Its maximum range of magnitudes (12.3–14.2) is well within the visual range of any experienced amateur with a 30-cm telescope, although blotting out that annoying 6th-magnitude star with an occulting bar in the eyepiece field makes it a lot easier. In professional circles Mrk 421 is well known as a TeV gamma-ray source, and there are tantalizing hints of a 23-year periodicity in its outbursts.

OJ 287

In recent years the blazar OJ 287 (see Figure 5.9) has been the focus of intense scrutiny by amateur astronomers with large telescopes or access to CCD equipment. Although its quiescent magnitude of 15–16 (and lower) makes it a challenging, or even impossible, target for many amateur astronomers, the nature of the object is intriguing. OJ 287 can vary in intensity on a timescale of tens of minutes, even in the optical waveband. In addition, it seems to have major outbursts, which indicates a period of 12 years or so.

At first, a periodic nature to any black hole-powered system might seem hard to comprehend; are we seeing matter falling into a black hole on a regular basis? In fact,

Figure 5.9. The extraordinary blazar OJ 287 appears very star-like in this image by the author. In fact it is thought to be an active galaxy with a binary black hole at its center. Imaged by the author on May 15, 2006, with a 0.35-m Celestron 14 at f/7.7 and SBIG ST9XE CCD. 120-second exposure. The field is 13 arcminutes square with north up, and the magnitude of OJ 287 was 15.8 at the time.

this is not the nature of the periodicity in such AGN outbursts. There are two categories of theory: namely, those that involve a binary supermassive black hole at the AGN center, and those that do not. Those that do not involve a binary black hole (BBH) invoke a quasi-periodic oscillation of the accretion disk surrounding the black hole or jet-induced outbursts. However, the BBH idea is quite popular, even if it sounds like an idea straight out of science fiction. In fact, when two galaxies have merged, simulations show that BBHs are not that rare. It all depends on the galaxy types and black hole mass ratios. Obviously, with such phenomena being the result of galaxy mergers, they would most likely be found in galaxy superclusters.

Theories suggest that the orbital periods of such BBHs would probably be on the order of tens of years. When proposing that a blazar has, say, a 12-year period, data is required going back well beyond the modern era; obviously such objects only appeared as variable stars prior to the 1960s. Fortunately, photographic plate archives such as those of the Smithsonian and Sonneborg observatories survive and are available for such data trawls.

In the case of OJ 287, various BBH models have been put forward to explain the possible 12-year optical outburst cycle (12 years with a double peak one year in width). The model of Dr Mauri Valtonen (using a simulation of 1 million particles in the accretion disk) has generated the most interest in recent years (see Figure 5.10a, b). In this model the outbursts are proposed to have been generated by impacts of the smaller (100 million solar masses) black hole on the larger (18 billion solar masses) black hole's accretion disk, precession of the relativistic jet, and the tidal effects of the smaller black hole on the larger black hole's accretion disk. The whole system is

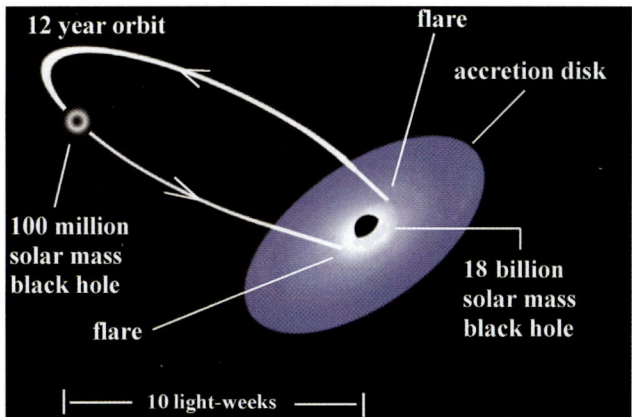

Figure 5.10a. The most likely orbital configuration of the binary black hole AGN called OJ 287 is shown in the diagram. A 100-million-solar-mass black hole orbits a 180× more massive 18-billion-solar-mass black hole every 12 years or so. The orbit of the smaller body is highly elliptical, with the monstrous larger black hole at one focus of the ellipse. As well as tidal variations in brightness and fluctuations caused by the volume of infalling matter, major outburst flares occur when the smaller black hole crashes through the accretion disk surrounding the larger one. The entire system is thought to span no more than a few light-months, maybe 1/20th of the distance from our Sun to the nearest star. Yet it contains over 18 billion solar masses! Diagram by the author based on the models of Mauri Valtonen.

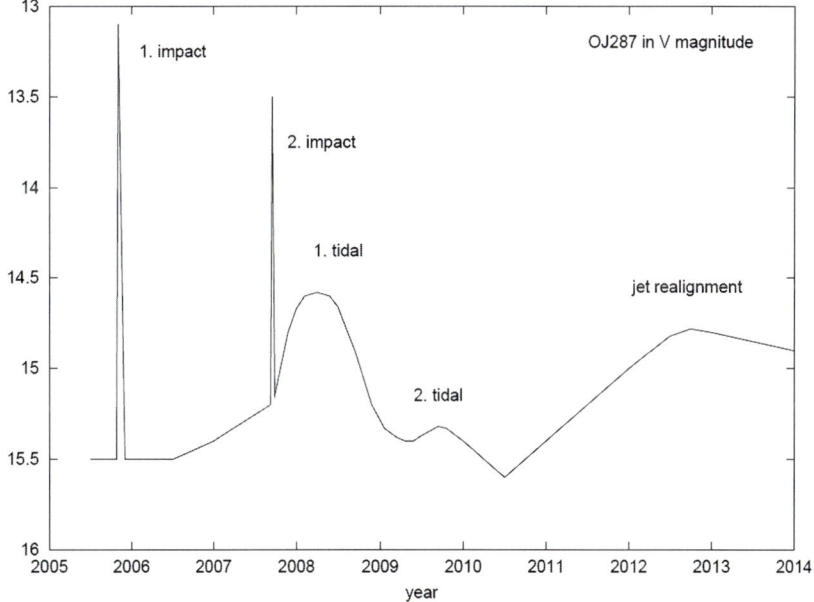

Figure 5.10b. Dr Mauri Valtonen predicted the long-term visual magnitude of OJ 287 as shown in the graph. The huge brightness spikes correspond to the impact times of the smaller black hole as it passes through and back through the larger black hole's accretion disk. The other variations are tidal (gravitational) in nature. The prediction, so far, has proved very accurate and is based on a 1-million-particle simulation. This prediction should be compared with the previous figure (Figure 5.10a) to visualize what is happening. Graph by kind permission of Dr Mauri Valtonen.

thought to fit into an area of space only light-months across. This theory is based on 500 data points from the historic 1890–1970 Sonneborg photographic archive as well as the mass of data from the past few decades, much of which is from the amateur astronomers equipped with CCDs from the 1990s onwards. In the 1990s OJ 287 flared to around magnitude 14.8 on December 11, 1993, but rapidly faded back to 15.6 by December 21. Three months later, on March 3, 1994, it flared to magnitude 14.5, but in only 2 days it fell back to 15.5. Outbursts occurred in October (magnitude 14.0 on October 10) and November 1994, roughly in accordance with predictions. Toward the end of 1997 a low state of magnitude 16 was observed again, roughly as predicted.

Valtonen predicted major outbursts of OJ 287 would occur in March 2006 and in September 2007, due to the fact that those dates would be the double peak. In fact, OJ 287 did have a peak around late March/early April 2006, at around magnitude 14.1, but it was actually higher in November 2005, when it peaked at magnitude 13.9.

From mid-June to early September each year OJ 287 is lost in twilight, and so for almost 3-months optical data is lost. However, in the first week of September 2007 observers recovered this peculiar object at magnitude 14.2. This was an increase of roughly 0.3 magnitudes since June and, although not a record, it showed that OJ 287 was technically 'in outburst.'

Throughout September and October 2007 the average visual magnitude of OJ 287 crept up, reaching 13.8 by late October. So this seemed to indicate that the autumn 2007 prediction was correct, even if the peak was not, perhaps, as clearcut and record breaking as has been hoped. The theory still needs refining, and this is one object where continued monitoring by amateurs over the longer term will reap rewards. There are not many BBH AGN candidates out there to study! A complete light-curve, compiled by Gary Poyner, appears in Figure 5.11. Remarkably, the observations of

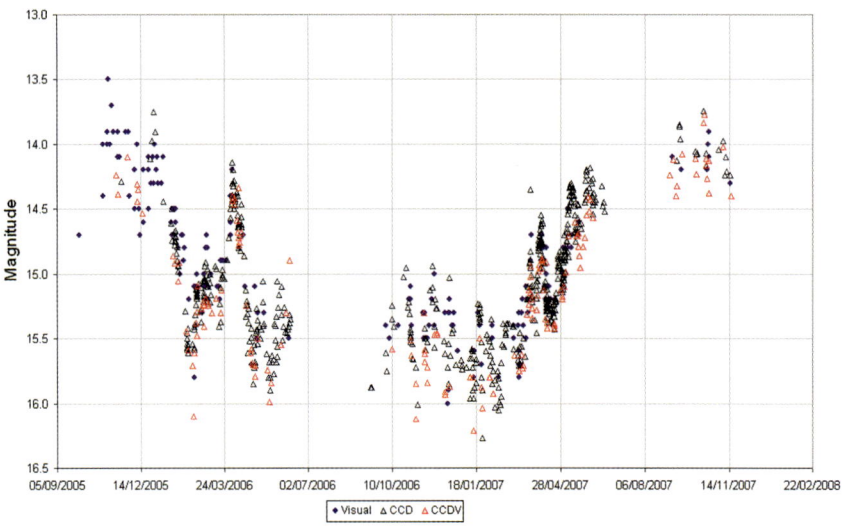

Figure 5.11. The light-curve of OJ 287, compiled by Gary Poyner, from observations submitted to the British Astronomical Association between September 2005 and the start of 2008. During that time the AGN varied between magnitudes 16.3 and 13.5, with the September/October peak agreeing well with predictions. Following the deluge of amateur results on this distant object, professional astronomers have been able to revise the main black hole's mass upwards from 17 billion solar masses to 18 billion solar masses! Graph by kind permission of Gary Poyner and the BAA variable star section.

amateurs have been so valuable that they have been used to revise upwards the estimated mass of the larger black hole in the OJ 287 system. Remember, a heavier body will affect the orbital characteristics of the smaller one, decreasing the orbital period. In 2007 the result of these amateur observations led OJ 287's parent black hole to be upgraded from 17 billion solar masses to 18 billion solar masses. Surely there has never been another case when amateur observations added so much weight to a cosmic body?

The Go To list of most telescopes will not feature OJ 287 in their database; however, slewing to galaxy IC 2423 and then moving another 7 arcminutes south will put OJ 287 in the center of the field.

3C 66A

Known as 3C 66A (see Figure 5.12) this distant BL Lac object is thought to have a supermassive black hole at its core, namely, one of around 25 million solar masses. It appears to have a 65-day periodicity in its light-curve, which may be the result of instability in the accretion disk surrounding the black hole. 3C 66A lies in the constellation of Andromeda (2 h 22m 40s +43° 2' 6''), within a few arcminutes of the faint galaxies UGC 1832, 1837, and 1841, and only two-thirds of a degree north of the well-known and bright edge-on galaxy NGC 891 (also known as Caldwell 23). It is a sobering thought that, at 31 million light-years, NGC 891 is roughly 130 times nearer than 3C 66A, which is thought to be 4 billion

Figure 5.12. The field of the active galaxy 3C 66A also contains the galaxies UGC 1841 (below bright star) and UGC 1837 (lower left) and UGC 1832 top right of center. 3C 66A itself is the anonymous looking 'star' in the dead center of this 15 arcminute field (north is up). Image credit: STScI/DSS.

light-years distant. This blazar varies between a maximum magnitude of 13.8 and a minimum magnitude of 15.6. In a typical year it fluctuates between 14.0 and 15.0. 3C 66A was in a high state for most of 1993 and right through to mid-1997. It then spent the best part of a year below magnitude 15, just prior to a steady rise to an absolute maximum of 13.8 at the end of 1998. In the northern hemisphere's spring of 2004 it once again peaked above magnitude 14: a blazar well worth monitoring if only for the novelty of observing an object that is 4 billion light-years distant from your backyard!

W Comae

Like BL Lac, W Comae (in Coma Berenices) was originally classified as a variable star, hence its designation. It has been observed ever since its variability was discovered on photographic plates in 1911 (by Max Wolf). At its position (12 h 21m 31.5s, +28° 13' 57") W Comae lies in northwestern Coma Berenices, not far from the southern border regions of Canes Venatica and Ursa Major. Find 4th-magnitude Gamma Comae, move just more than 1 degree due west, and there is W Comae, 6 arcminutes northeast of galaxy NGC 4295 and less than an arcminute northwest of another 12th-magnitude star, a real star in this case. W Comae (also called ON 231) is another blazar and is regarded as showing the most intense gamma-ray spectrum of all 60 known gamma-ray blazars. Major outbursts in brightness have occurred in 1940, 1952, 1968, and 1980. The British amateur astronomer Gary Poyner discovered further outbursts, visually, in 1995 and 1998. The latter outburst saw W Comae peak at magnitude 13.2. Normally it stays between 14.0 and 15.0, although fades to magnitude 15.5 and even fainter are possible.

S5 0716+71

This rather painfully named blazar is yet another one that lies in the far northern hemisphere. At 7 h 21m 53.4s and +71° 20' 36", it lies in Camelopardalis and is circumpolar from Europe and North America, so it can be monitored all year round. The object normally varies between magnitudes 13.0 and 15.0, but at the start of 2004 it reached an all-time high of around magnitude 12.5, and a low of 15.5 is possible. Variations of more than 1 magnitude in a month are rare, but experienced amateur astronomers with CCD cameras can often detect smaller hourly variations of 0.1 magnitudes or more.

Seyfert Galaxies

Seyfert galaxies are the closest category of AGN objects to us. The examples that amateurs can study are not at the cosmological distances of the quasars, and they do not have a jet pointed directly at us. Because of their proximity they can be seen or imaged as obvious galaxies with amateur equipment, rather than just dots. Seyferts are named after Carl Keenan Seyfert, who was the first to identify this category of galaxy in the 1940s. Unlike the quasars there was no astonishment when the distance of Seyferts was measured, as they are, relatively speaking, local objects.

Figure 5.13. This excellent Hubble Space Telescope image of the Seyfert galaxy NGC 7742 illustrates the classic appearance of this type of active galaxy when viewed face-on. An intense central region is surrounded by a thick 'lumpy' ring, 3000 light-years from the core. The ring is an area of active starbirth. Credit: Hubble Heritage Team (AURA/STScI/NASA/ESA).

Seyferts are currently classified using a fractional notation based on the relative strength of the narrow and broad components in the spectrum. Both components are thought to originate from the accretion disk surrounding the central black hole. A Type 1.0 Seyfert would show clear narrow and broad emission lines, but a Type 2.0 Seyfert would only show narrow lines, due to (presumably) dust or our viewing angle obscuring the broad emission lines. In between these two (original) types are Types of 1.1–1.9, depending on the relative amount of narrow and broad emission lines. Seyfert galaxies have unusually bright cores, and are almost stellar in appearance (see Figure 5.13). The two best examples that amateurs can study are described below.

M77

It may sound absurd, but not only is the nucleus of the active and well-known galaxy Messier 77 (M77) (see Figure 5.14) variable in brightness, it is poorly observed, too. Now do not misunderstand. M77 itself is not poorly observed. As one of Messier's objects it is a very popular deep-sky object in the autumn sky, and being so close to the celestial equator, it is visible from both hemispheres. M77 lies in the northernmost part of Cetus the Whale, where the whale's head just fits between Pisces and Taurus. Variable star observers will know the region well, as M77 is just 5 degrees northeast of the long-period red giant star Mira (Omicron Ceti). However, very few observers treat M77 as a variable star!

Figure 5.14. The central regions of the best-known Seyfert galaxy, M77 are clearly shown in this image by the author. Taken on January 21, 2007, with a 0.35-m Celestron 14 @ f/7.7 and SBIG ST9XE CCD. 2×180 seconds.

Of course, an object that has a definite size in an amateur telescope can be tricky to allot a magnitude to. As with comets, certain techniques need to be adopted to assess the brightness of a fuzzy object. These techniques include using a smaller telescope to make the object more starlike, to allow nearby comparison stars to enter the field, and to defocus the wide field until stars appear as fuzzy as the galaxy. The photometric CCD approach is to use an 'aperture,' or radius, in arcseconds that just includes the region of interest and no more. In long exposures of M77 the nucleus is easily overexposed in the standard 'contrast-stretch' operation. However, the central part is almost stellar in appearance in optimally processed images.

M77 is also known as NGC 1068, 3C 71, and Cetus A. The latter two designations are due to its strong radio emissions. Experienced observers suspect that the central region of M77 varies in brightness by a magnitude or more, but it is rarely treated as a variable object, so maybe a reader of this book might like to take up the challenge. As the second nearest Seyfert galaxy, at just 60 million light-years away, it is surely the easiest AGN to study visually with amateur equipment. Prior to the fractional notation for Seyferts, M77 was listed as a Seyfert 2 object, until Antonucci and Miller (1985) found that it showed broad emission lines, too, in plane-polarized light.

NGC 4151

At a distance of around 42 million light-years the galaxy NGC 4151 (Figures 5.15 and 5.16) is the closest Seyfert, even beating M77 in that respect. Like so many good AGNs it resides in the far north of the sky, specifically, near the western Canes Venatica border with Ursa Major, at 12 h 10m 1.7s, 39° 52′ 55″. The nucleus of NGC 4151 was unusually bright between 1989 and 1997, with notable maxima in 1993 and 1995.

Figure 5.15. This figure shows four images, at different wavelengths, of the Seyfert galaxy NGC 4151, imaged by the Hubble Space Telescope. The top left image shows the galaxy at the wavelength of the 5007 Å oxygen emission line. The lower left image is from the HST's spectrograph and shows the velocity distribution of the carbon emission from the gas in the core of NGC 4151. The top right image shows the spectrum of the oxygen gas emission. The bottom right image is a false-color image of the two emission lines of oxygen gas at 4959 Å and 5007 Å. Image STScI/NASA.

Unlike M77, NGC 4151 has been studied in detail by a few amateur astronomers and by professionals too. Typically, professionals use tiny CCD measurement apertures of 15–20 arcseconds on the sky to monitor NGC 4151 in UBVRI wavelengths. The central black hole of NGC 4151 is thought to have a mass of 10 billion solar masses; a figure derived from observations by the International Ultraviolet Explorer's study of the gas surrounding the inner nucleus. In the visual waveband NGC 4151's nuclear region varies between magnitudes 11 and 13. However, as we just noted with M77, estimating the magnitude of a faint fuzz, nestling within an even bigger fuzz, visually, is a major challenge, even for experienced visual observers. With a large amateur telescope the fainter 13th-magnitude-galaxy NGC 4156 may be spotted some 4 arcminutes to the northeast of NGC 4151.

Figure 5.16. An optical image of the same Seyfert galaxy shown in the previous figure, namely NGC 4151. North is up, and the field is 15' wide. Credit: STScI/DSS.

Active Galaxies for Observing

Table 5.2 summarizes which active galaxies should be considered for observation.

As well as monitoring AGNs for variability, many of them make challenging deep-sky imaging targets too. This is especially so for the cases of the Virgo galaxy M87 and, specifically, its jet, and the bizarre radio galaxy Centaurus A, visible from the southern hemisphere. Various features in the field of the active galaxy NGC 1275 (Perseus A) are shown in Figure 5.17.

Table 5.2. Sixty-five active galaxies that can be monitored visually or with CCDs by amateur astronomers

Name	R.A. (2000.0)	Dec. (2000.0)	Mag. range	Type	CON	Redshift
3C 66A	02 h 22m 39.61s	+43° 02' 07.8"	13.5–16.0	BLAZ	And	0.444
3C 147	05 h 42m 36.14s	+49° 51' 07.2"	17–18?	SEYF	Aur	0.545
3C 232	09 h 58m 20.95s	+32° 24' 02.2"	15.3–16.2	SEYF	Leo	0.530
3C 273	12 h 29m 06.70s	+02° 03' 08.6"	12.0–13.4	QSO	Vir	0.158
3C 279	12 h 56m 11.17s	–05° 47' 21.5"	13.3–16.3	BLAZ	Vir	0.536
3C 345	16 h 42m 58.81s	+39° 48' 37.0"	15.5–<16	BLAZ	Her	0.593

(Continued)

Name	R.A. (2000.0)	Dec. (2000.0)	Mag. range	Type	CON	Redshift
3C 351	17 h 04m 41.38s	+60° 44′ 30.5″	15–16?	SEYF	Dra	0.372
3C 371	18 h 06m 50.68s	+69° 49′ 28.1″	13.1–15.9	BLAZ	Dra	0.051
3C 382	18 h 35m 03.39s	+32° 41′ 46.9″	12.5–14.5	SEYF	Lyr	0.058
3C 454.3	22 h 53m 57.75s	+16° 08′ 53.6″	16.0–16.5	QSO	Peg	0.859
4C 29.45	11 h 59m 31.83s	+29° 14′ 43.8″	15.6–18.1	QSO	UMa	0.729
AKN 120	05 h 16m 11.48s	–00° 09′ 00.6″	13.9–14.1	SEYF	Ori	0.034
AO 0235+16	02 h 38m 38.93s	+16° 36′ 59.3″	15.5–16.5	BLAZ	Ari	0.940
Mrk 205[1]	12 h 21m 44.12s	+75° 18′ 38.2″	13.9–15.2	SEYF	Dra	0.070
Mrk 335	00 h 06m 19.52s	+20° 12′ 10.5″	13.7–13.9	SEYF	Peg	0.025
Mrk 421	11 h 04m 27.31s	+38° 12′ 31.8″	12.5–14.3	BLAZ	UMa	0.030
Mrk 478	14 h 42m 07.46s	+35° 26′ 22.9″	~14.5	SEYF	Boo	0.077
Mrk 501	16 h 53m 52.22s	+39° 45′ 36.6″	13.5–14.0	BLAZ	Her	0.034
Mrk 530	23 h 18m 56.61s	+00° 14′ 36.5″	~14.0	SEYF	Psc	0.029
Mrk 509	20 h 44m 09.76s	–10° 43′ 24.7″	~13.0	SEYF	Aqr	0.034
Mrk 590	02 h 14m 33.60s	–00° 46′ 00.3″	~14.0	SEYF	Cet	0.027
Mrk 1502	00 h 53m 35.08s	+12° 41′ 34.4″	~14.0	SEYF	Psc	0.061
OJ 287	08 h 54m 48.88s	+20° 06′ 30.6″	13.9–16.6	BLAZ	Cnc	0.306
PKS 0003+15	00 h 05m 59.24s	+16° 09′ 48.9″	~16.5	SEYF	Peg	0.450
PKS 0716+71	07 h 21m 53.45s	+71° 20′ 36.4″	12.3–14.8	BLAZ	Cam	0.300
PKS 0735+178	07 h 38m 07.39s	+17° 42′ 19.0″	14.6–16.4	QSO	Gem	0.424
PKS 0736+01	07 h 39m 18.03s	+01° 37′ 04.6″	16.7–17.0	QSO	CMi	0.191
PKS 1354+195	13 h 57m 04.44s	+19° 19′ 07.4″	~16.0	SEYF	Boo	0.719
PKS 1622–29	16 h 26m 06.02s	–29° 51′ 27.0″	15.3–15.9	QSO	Sco	0.815
PKS 1749+096	17 h 51m 32.80s	+09° 39′ 02″	14.1–18.4	QSO	Oph	0.322
PKS 2155–30	21 h 58m 52.07s	–30° 13′ 32.1″	12.3–13.9	BLAZ	PsA	0.117
S2 0109+224	01 h 12m 05.83s	+22° 44′ 38.8″	14.8–15.5	BLAZ	Psc	TBD
S4 1749+70	17 h 48m 32.84s	+70° 05′ 50.8″	~16.7	QSO	Dra	0.770
S5 1803+78	18 h 00m 45.68s	+78° 28′ 04.0″	14.0–16.5	BLAZ	Dra	0.684
S5 2007+77[2]	20 h 05m 30.99s	+77° 52′ 43.3″	~16.5	BLAZ	Dra	0.342
IO And	00 h 48m 18.98s	+39° 41′ 11.6″	15.3–17.6	SEYF	And	0.134
S10721 And	00 h 38m 33.10s	+41° 28′ 50.0″	16.9–19.0	SEYF	And	0.0725
S10838 Aur	05 h 54m 53.60s	+46° 26′ 22.0″	14.4–15.5	SEYF	Aur	0.0205
CC Boo	13 h 40m 22.84s	+27° 40′ 58.6″	17.8–19.5	SEYF	Boo	0.172
CD Boo	13 h 41m 23.34s	+27° 49′ 55.4″	18.9–19.7	QSO	Boo	1.045
S10765 Boo	13 h 46m 47.2s	+29° 54′ 20″	17.0–18.8	SEYF	Boo	0.063
XX Cet	02 h 22m 39.90s	–19° 32′ 47.0″	18.0–19.7	QSO	Cet	0.736
GQ Com	12 h 04m 42.11s	+27° 54′ 11.5″	14.7–16.1	SEYF	Com	0.165
W Com	12 h 21m 31.69s	+28° 13′ 58.5″	11.5–17.5	BLAZ	Com	0.102
X Com	13 h 00m 22.14s	+28° 24′ 02.5″	12.5–17.9	SEYF	Com	0.092
AU CVn	13 h 10m 28.66s	+32° 20′ 43.8″	14.2–20.0	QSO	CVn	0.996
S10764 CVn	13 h 42m 10.90s	+28° 28′ 48.0″	18.3–19.8	QSO	CVn	0.330
V1102 Cyg	19 h 10m 37.07s	+52° 13′ 14.9″	15.5–17.0	SEYF	Cyg	0.027
V395 Her	17 h 22m 34.00s	+24° 45′ 01.0″	16.1–17.7	SEYF	Her	0.0638
V396 Her	17 h 22m 43.20s	+24° 36′ 18.0″	15.7–16.7	SEYF	Her	0.175

(*Continued*)

Chapter 6

Gamma Ray Bursters

Gamma-ray bursters, or GRBs, are, quite simply, the most energetic events in the universe. The amounts of energy involved in these cosmic explosions are, even by cosmological standards, simply staggering. This whole book is about unbelievably high-energy events occurring in our galaxy and in distant and superdistant galaxies. But even the cosmologically remote quasars do not emit brief bursts of energy that stand in comparison with the most dramatic GRBs. The GRB from a nearby supernova explosion has been suggested as the cause of the Ordovician extinction, 450 million years ago, which resulted in the death of nearly 60 percent of the oceanic life on this planet.

The first question a newcomer to GRBs will undoubtedly ask is: "How can an amateur detect gamma rays?" Fortunately, they do not have to, as GRBs often produce an optical counterpart, too, and it is detecting and monitoring the fade of the optical counterpart that is the vital amateur contribution.

GRBs were first detected by the US's Vela satellites in the 1960s, which were designed to detect secret nuclear weapons tests by using onboard X-ray, neutron, and gamma-ray detectors. The satellites never positively detected any such nuclear tests, with one exception. On September 22, 1979, the Vela 6911 satellite detected a possible nuclear explosion in the Indian Ocean, roughly midway between South Africa and the Antarctic. Although this was never confirmed as a nuclear test, there were strong suspicions that it was a device detonated by South Africa. However, the satellite was, by then, 2 years beyond its predicted working lifetime.

Even though the Vela satellites detected no confirmed nuclear tests, they did detect mysterious bursts of gamma radiation coming from space. The first of these bursts appears to have occurred on July 2, 1967. In June 1973 the *Astrophysical Journal* (Vol 182) published a paper by Klebesadel, Strong, and Olson entitled 'Observations of gamma-ray bursts of cosmic origin,' in which they analyzed 16 energy bursts in the 0.2–1.5-MeV energy range detected between July 1969 and July 1972, an era mainly remembered for the Apollo moon landings. That paper started the search to find out where in deep space the GRBs were coming from. However, it was not until 1991 that astronomers were able to start to pin down the origin of GRBs, other than, that they came from space.

The year 1991 saw the launch of the Compton Gamma Ray Observatory satellite and, crucially, its Burst and Transient Source Explorer (BATSE) gamma-ray detector. BATSE was able to show that GRBs occurred all around the sky and were not confined to our Solar System, or even the Milky Way Galaxy. This was a staggering result, as an extragalactic origin implied that GRBs were very distant and therefore very powerful. BATSE also showed that, in general, there appeared to be two types of GRBs: short, intense bursts of less than 2-seconds duration and longer bursts of lower intensity.

M. Mobberley, *Cataclysmic Cosmic Events and How to Observe Them*,
DOI: 10.1007/978-0-387-79946-9_6, © Springer Science+Business Media, LLC 2009

However, what astronomers really wanted was to be able to pin down precisely the position of a GRB and look at the object optically. Early gamma-ray detectors had very poor angular resolution, simply because of the difficulty of focusing radiation that, unlike light, tends to pass through the instrument measuring it! Still, with increasingly sophisticated detectors, the resolution increased to the point where astronomers hoped that an approximate location of the GRB source might enable the field to be scoured optically and a fading counterpart to be detected with conventional telescopes.

What Causes GRBs?

The amount of energy released in GRB events is truly staggering, and so cosmologists have had their work cut out trying to fathom mechanisms that might explain such a large energy release in such a short period of time. Even by the standards of supernovae and active galactic nuclei the figures involved would seem absurd, even in a science fiction novel. The most energetic GRBs have released an estimated 10^{47} joules of energy, if one assumes the energy is emitted uniformly in all directions. From Einstein's famous $E = mc^2$ formula, if we substitute the mass of the Sun ($m = 2 \times 10^{30}$ kg) and the speed of light ($c = 3 \times 10^8$ m/s), we get 1.8×10^{47} joules. So the most energetic GRBs are converting about 55 percent of the solar mass directly into its energy equivalent.

If you want a kinetic energy equivalence ($\frac{1}{2} mv^2$), it is a similar amount of kinetic energy to that released by the mass of our Sun, traveling at near-light speed, slamming into a brick wall (ignoring relativistic considerations). In fact, the most powerful GRBs emit as much energy in a matter of seconds as 1000 stars, like our Sun, emit over their entire lifetimes! In terms of power output (i.e., watts, or joules per second) this means that, at its peak, the power output of the most energetic GRBs is of a similar magnitude to the power output of every star in the visible universe.

It was hardly surprising, then, that astronomers rather wished that GRBs lived within our galaxy when trying to account for their energy levels; if they were a million-fold closer the power outputs required to explain the bursts decrease by a trillion-fold. However, since the detection of GRB 970228, that hope has disappeared. GRBs do occur billions of light-years away, so an ultrapowerful mechanism has had to be devised. Nevertheless, there is a 'get-out clause' in the 10^{47} joule (worst case) requirement, as it assumes the energy we see is emitted uniformly in all directions, throughout a sphere of surface area $4\pi r^2$, where r is the radius from the quasar to us. What if we are looking into a jet of energy, just like we are in the case of the active galactic nucleus of a blazar? Then the energy requirement drops considerably. For example, if the beam cone spanned a solid angle of 10 degrees, the energy requirement is reduced by a factor of (roughly) 500 and then comes within the range of the most violent supermassive supernova explosions, known as hypernovae.

Of course, bringing this factor of 500 (or so) into the equation has another effect, namely, it means there are probably 500 times as many GRB events that we do not detect. Is this a problem? Cosmologists think not, as the visible universe is a very large place containing well in excess of 100 billion galaxies, with each galaxy containing, say, 100 billion stars. If the mechanism for creating a GRB is an extremely rare form of supernova, there are still enough galaxies to provide a regular supply, even if we only see a tiny fraction of the events. Some GRB optical

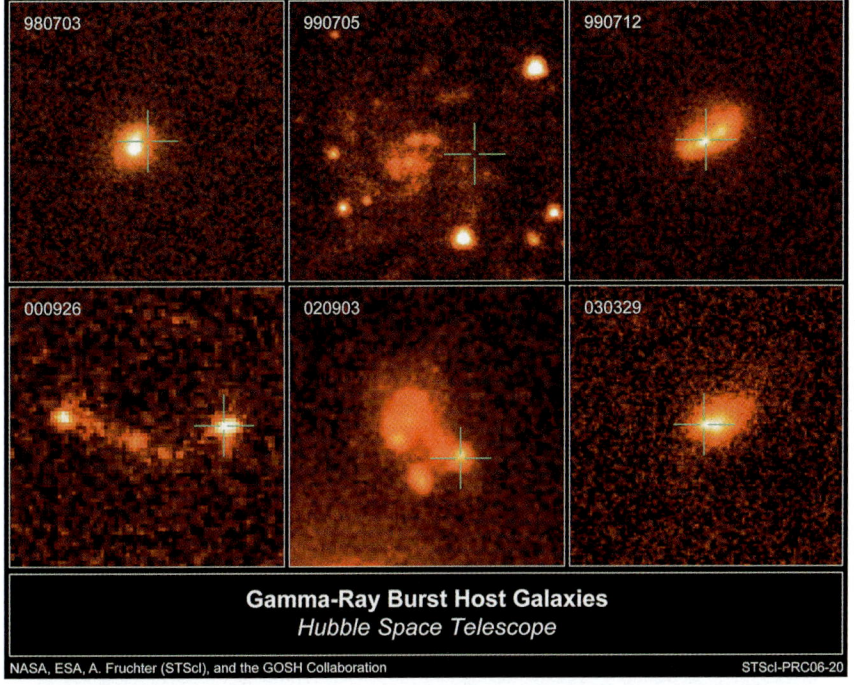

Gamma-Ray Burst Host Galaxies
Hubble Space Telescope

NASA, ESA, A. Fruchter (STScI), and the GOSH Collaboration   STScI-PRC06-20

Figure 6.1. Some famous gamma-ray bursts and their host galaxies, imaged by the Hubble Space Telescope. Credit: STScI/NASA/ESA/GOSH.

transients/supernovae and their distant host galaxies, imaged by the Hubble Space Telescope, are shown in Figure 6.1.

The current thinking is that the long-duration GRBs are caused by the supernova collapse of very massive stars, for example, stars with more than 40 times the mass of our own Sun. This theory is supported by the evidence that the fading optical afterglow of these events can sometimes be traced to vigorous star formation regions in distant spiral and irregular galaxies, where short-lived massive stars will be more abundant. There have also been a number of cases where GRBs have been detected a few days before an optical supernova has been observed in the same distant galaxy. It is even possible that a supermassive star can collapse so dramatically in a Type Ib/c supernova explosion that the black hole formed can swallow the entire star before the shock wave reaches the massive star's surface. Thus, to an onlooker, the supernova phase would not be seen, and only the GRB energy would arrive on Earth.

Short-Duration GRBs

As well as long-duration GRBs there are short-duration ones, too (see Figure 6.2). Short-duration GRB events are defined as those in which the burst of gamma rays last less than 2 seconds. In practice, the shortest bursts detected are only a few milliseconds long, but the typical short burst is between a quarter and half a second. By implication the longer bursts associated with the collapse of massive

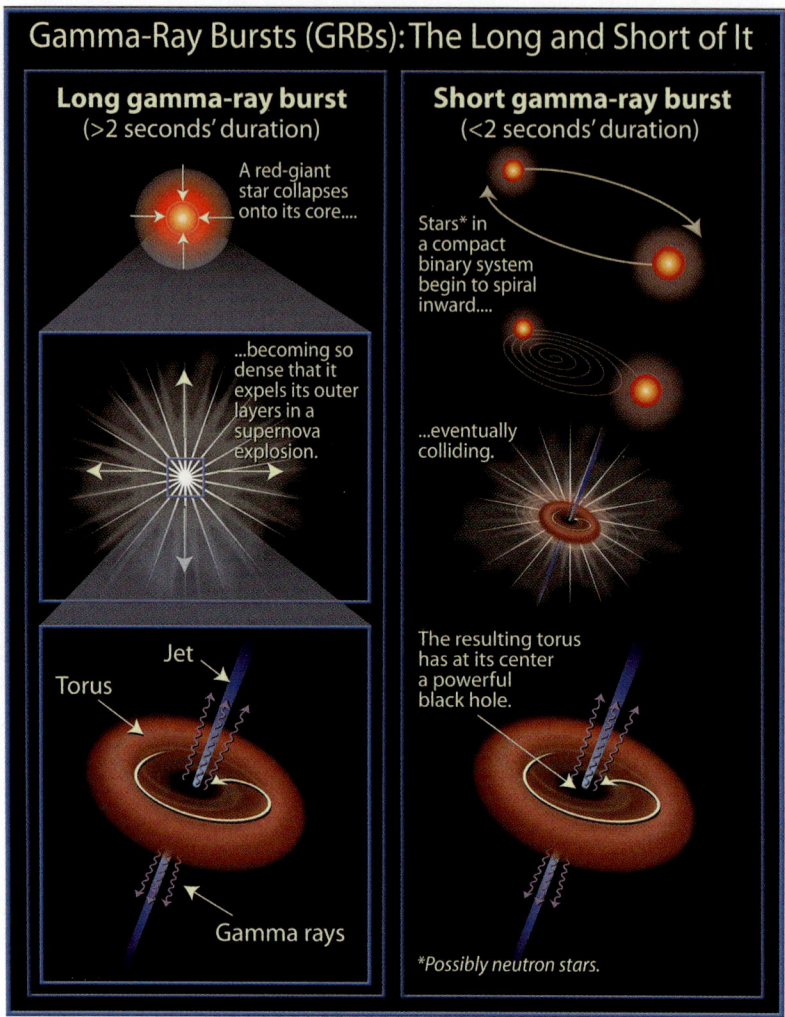

Gamma-Ray Bursts (GRBs): The Long and Short of It

Long gamma-ray burst
(>2 seconds' duration)

A red-giant star collapses onto its core....

...becoming so dense that it expels its outer layers in a supernova explosion.

Torus

Jet

Gamma rays

Short gamma-ray burst
(<2 seconds' duration)

Stars* in a compact binary system begin to spiral inward....

...eventually colliding.

The resulting torus has at its center a powerful black hole.

Possibly neutron stars.

Figure 6.2. Long and short gamma-ray burst mechanisms. Credit: NASA/ESA and Ann Field (Space Telescope Science Institute).

stars feature bursts with a typical length of 30 seconds, with the longest events being several minutes in duration.

The sub-2-second short GRBs are generally found in nearer, lower redshift galaxies (where a host galaxy has been identified), but not usually in the intense star-forming regions where one might expect massive short-lived stars to exist in abundance. Crucially, to date, no supernova has yet been linked to a short-GRB event. Astronomers have therefore looked toward alternative explanations for these high-energy events. The favorite theory for explaining these GRBs is the merger of two superdense objects, which have gradually spiraled into each other as predicted by relativistic gravitational radiation – specifically, two neutron stars, or a neutron star and a black hole.

In the last moments of the two neutron stars their mutual gravitation breaks the superdense objects apart, and astronomers think they then merge into a black hole with an intense burst of gamma radiation. Where one of the objects is a black hole already, it swallows the neutron star, again, with a huge burst of gamma radiation. Yet again, the likelihood of witnessing such an event from Earth is a numbers game: big numbers and tiny fractions. Binary neutron stars will be very rare in any galaxy, and the same holds for black hole–neutron star pairs. In addition, they will exist happily for millions of years before the cataclysmic split-second when they merge. However, offsetting that rarity, yet again, are the 100 billion plus galaxies in our observable universe. Somewhere out there, every day, neutron stars are colliding and black holes are swallowing neutron stars. In addition, the events are so colossal that the energy can be seen even by amateur astronomers billions of years and billions of light-years later, in the case of the biggest, long-duration GRBs.

Another option for the theorists, which may explain some short-duration GRBs, involves a special category of neutron stars called magnetars. Magnetars are neutron stars with a very powerful magnetic field; occasionally these release very high-energy flares and numerous short-duration bursts of hard X-rays (i.e., soft gamma-ray repeaters, or SGRs).

On December 27, 2004, a magnetar in our own galaxy (located in Sagittarius), some 50,000 light-years away, emitted an estimated 10^{39} joules of energy in 0.2 seconds – as much energy as the Sun radiates in a quarter of a million years. This was way below the estimated output of the most violent GRBs, but being 100,000 times nearer to Earth made it the most violent event recorded by gamma-ray detectors. It was probably the most violent event in our galaxy witnessed by humans since the supernova of 1604, although, if we had operated gamma-ray or X-ray detectors since that time we would undoubtedly have detected other, similar events.

This particular neutron star, cataloged as SGR 1806–1820, has a diameter of at most 20 km, a mass similar to that of our Sun, and completes one rotation every 7.5 seconds. As we have seen, the abbreviation SGR means soft gamma ray repeater, which can cause confusion, as Sgr is the astronomical abbreviation for Sagittarius, and, in this case, the event occurred in that constellation! SGR 1806–1820 was already known to astronomers as a soft gamma- and X-ray repeater and had already been described as the most powerful magnet in our galaxy, capable of stopping a metal locomotive at the distance of the Moon! But known or not this particular outburst was exceptional. In fact, although most of the energy was released in 0.2 seconds at 21h 30m 26.6s U.T. on December 27, the total outburst was 380 seconds in duration. Had the event not occurred within 2 days of the Asian tsunami (Dec. 26, 00:58:53 U.T.) it might have received considerably more attention.

Amateur astronomers made a contribution to the SGR 1806–1820 event despite the lack of a fading optical afterglow. The American Association of Variable Star Observers (AAVSO) submitted a report to the GRB coordination network (GCN), detailing how their members monitored a sudden ionospheric disturbance (SID) in Earth's atmosphere as a result of SGR 1806–1820. This manifested itself as a change in the signal strength from very low frequency (VLF) radio transmitters being monitored by stations around the world. The ionospheric disturbance was created by X-rays from the magnetar ionizing the upper atmosphere and adjusting its radio propagation characteristics. For more on VLF monitoring and SIDs, see Chap. 3.

In June 2006, a GRB was detected that refused to fit into either the long- or short-duration GRB category, just as professional astronomers thought they could slot everything into two distinct boxes! GRB 060614 burst in the high-energy domain for 102 seconds; so naturally, it qualified as a long-duration event. However, the manner in which the light faded was typical of a short-burst GRB. Currently the best possible explanation for this phenomenon is that we were not witnessing two neutron stars, or a neutron star and a black hole merging; instead, we were seeing a neutron star merging with a white dwarf. In reality, the jury is still out on this one.

Some Historic GRBs

GRB 970228

R.A. 5h 01m 46.70s Dec. +11° 46′ 53.0″
Peak magnitude: Unknown
Magnitude 1 day later: ~21

The Italian/Dutch satellite BeppoSAX, launched on April 30, 1996, contained five X-ray detectors that, when used together, held the promise of pinpointing a GRB outburst with sufficient precision for an optical identification. This dream was realized on February 28, 1997, when GRB 970228 was detected, and its Wide Field Camera derived a preliminary position in northwestern Orion, roughly midway between Orion's Bellatrix and the bright star Aldeberan in Taurus.

When the 4.2-m William Herschel telescope was swung onto the field, a starlike object, fading from magnitude 21.3 to below magnitude 23.6 (between March 1.0

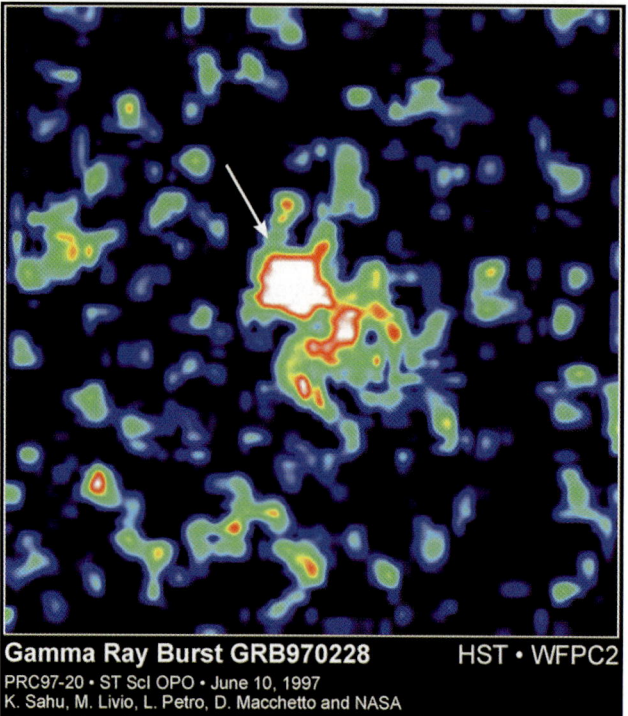

Gamma Ray Burst GRB970228 **HST • WFPC2**
PRC97-20 • ST ScI OPO • June 10, 1997
K. Sahu, M. Livio, L. Petro, D. Macchetto and NASA

Figure 6.3. A false-color Hubble Space Telescope image of the fading afterglow of GRB 970228 and its host galaxy. Credit: NASA/STScI.

and March 8.9) was located only 3 arcminutes from the BeppoSAX position, that is, at 5h 01m 46.70s and +11° 46′ 53.0″. Not surprisingly the Hubble Space Telescope was turned onto the field and by March 25 the fading GRB afterglow had dropped to magnitude 25.7 in the V (visual) band. The event seemed to have occurred (Figure 6.3) in a host galaxy with a redshift, z, of 0.7, indicating a distance of 6 billion light-years and a colossal energy output. At last this was solid evidence that GRBs were probably incomprehensibly energetic events at truly cosmological distances. It also proved that swinging major telescopes onto the likely search area, once a GRB is detected, could pay big dividends.

GRB 970508

R.A. 6h 53m 49.20s Dec. +79° 16′ 19.0″
Peak magnitude: Unknown
Magnitude 1 day later: $\sim$21

This GRB is regarded as the first one that unequivocally proved that these objects are extragalactic, as the redshift of the fading afterglow (not the host galaxy) was measured directly. Even at a magnitude of 21 the GRB greatly outshone its host, a 25th-magnitude dwarf galaxy only 10,000 light-years or so in diameter, similar to our own Large Magellanic Cloud (LMC). Optical spectra of the 21st-magnitude GRB afterglow was secured by the giant 10-m Keck telescope, which revealed both absorption and emission lines at a redshift of $z = 0.835$, indicating a distance of almost 7 billion light-years. Even a month after the outburst the GRB outshone its host galaxy.

GRB 971214

R.A. 11h 56m 26.40s Dec. +65° 12′ 00.5″
Peak magnitude: Unknown
Magnitude 1 day later: $\sim$21

GRB 971214 triggered all of the world's GRB detectors when it burst forth on December 14, 1997. The fading I band and X-ray afterglows and the host galaxy of the GRB were observed, using telescopes at Kiso (Japan), Kitt Peak (Arizona), the BeppoSAX satellite, and the giant 10-m Keck II telescope (Hawaii). The host galaxy was found to lie at a redshift of $z = 3.4$, implying a distance to the GRB of a massive 14 billion light-years. At the time of the outburst, which was massive by any normal standards, some rather overhyped estimates of the energy output of GRB 971214 were made in the press, which failed to take into account that the energy was probably fortuitously 'beamed' in our direction rather than spread over a 360-degree sphere of the sky. Frankly, it did not need overhyping!

GRB 980425 = SN 1998bw

R.A. 19h 35m 03.17s Dec. −52° 50′ 46.1″
Peak magnitude: 14? (supernova mag)
Magnitude 1 day later $\sim$14 (supernova mag)

This was the first GRB to be associated with a supernova; in this case supernova 1998bw in the far southern hemisphere galaxy of ESO 184-G82. The supernova was located in a spiral arm of that barred spiral galaxy, and the small redshift of ESO 184-G82, a z of a mere 0.0085, placed the galaxy little more than 100 million light-years away. It was easily the closest GRB ever detected. At 14th-magnitude SN 1998bw was an easy amateur target, too, provided you lived in the southern hemisphere. The best current guess is that SN 1998bw was a collapsar, a rapidly rotating giant Wolf–Rayet star with a mass of more than 30 times that of our own Sun. As we have already seen in this book this type of supernova/hypernova collapses directly into being a black hole, upon which two energetic plasma jets emerge from its rotational poles and emit intense gamma rays, hence the GRB pulse. Officially the supernova was classified as Type Ic due to its radio spectrum.

Prior to this crucial GRB/supernova tie-up there had been other supernovae suspected of being associated with GRBs, but the evidence had simply not been strong enough. For example, GRB 970514 and SN 1997cy had been suggested as being linked, and a connection between GRB 971115 and SN 1997ef was strongly suggested, too. Shortly after 1998bw another Type Ic 'collapsar,' 1998ey, with a very similar spectrum to SN 1997ef, was discovered, and yet there was no GRB event to tie up with that event. However, 1997ef and 1998ey were far less powerful supernovae, and, despite some astronomers starting to doubt a GRB/SN link, the GRB 980425/SN 1998bw connection finally laid the doubts to rest. These two phenomena were most definitely the same colossal event: the collapse of a massive star.

GRB 990123

R.A. 15h 25m 30.10s Dec. +44° 46′ 00.0″
Peak magnitude: 9!
Magnitude 1 day later: 20

On January 23, 1999, the Compton gamma ray observatory's BATSE detector recorded another energetic GRB. The approximate location was relayed to the GCN located at NASA's Goddard Space Flight Center in Greenbelt, Maryland. By this time astronomers had fine-tuned the reporting chain so that the period from detection to swinging an optical instrument onto that patch of sky was minimized. Within a mere 22 seconds of the GRB alert being forwarded by GCN to participating observatories, the Robotic Optical Transient Search Experiment (ROTSE I) instrument at Los Alamos, New Mexico, started imaging the field. This system (now superceded) used four commercial 200-mm f/1.8 Canon lenses, with cooled CCD detectors, to cover a 16×16-degree field of view down to magnitude 15. The compact system was housed on the roof of a small, protective shelter, a bit like a white garden shed, with a rapidly opening hinged top, for swift access to the sky.

Five seconds after ROTSE 1 was trained on the field of GRB 970123 the object peaked in activity as a 9th-magnitude object in the ROTSE field. Incredibly, it would have been visible in binoculars if anyone had been looking in that position. For an object so far away to emit so much radiation the energy released, even by GRB standards, must have been absolutely colossal. Seven minutes later it had faded from 9th to 14th magnitude. Four hours after the GRBs detection the BeppoSAX satellite pinned the object's position down to a few minutes of arc, enabling the historic 60-inch (1.52 m) Mt Wilson telescope to image the 18th-magnitude fading remnant. The next day the 10-m Keck II telescope was used to obtain a spectrum of the GRBs host galaxy. It turned out to be a staggering 9 billion light-years away! Two weeks later, on February 8/9, the Hubble Space Telescope was used to image the fading GRB source, now a 25th-magnitude object. For an object at 9 billion light-years to shine at 9th magnitude means that its absolute magnitude (i.e., at 10 parsecs, or 32.6 light-years) would be –33! Put another way, even from a distance of 500 light-years it would briefly have exceeded the brightness of our Sun in our daytime sky. It is like 20 times the output of the most active galactic nucleus known, just flaring up, from nothing, for a matter of seconds. If you want another comparison, GRB 990123 briefly shone with the light of a million billion stars like our Sun!

A New GRB Detection Era

Sadly, on June 4, 2000, after 9 years of service, the Compton gamma-ray observatory satellite was de-orbited due to the failure of one of its gyros. This was a somewhat controversial decision, as the CGRO instruments covered an unprecedented six decades of the electromagnetic spectrum, from 30 keV to 30 GeV, and its BATSE detector had been the major GRB event-detection system. BeppoSax was still in orbit, but its main role had been X-ray astronomy and pinpointing GRB positions. Thus, something of a lull occurred in GRB alerts until the next generation of satellites were launched, namely HETE-2 in October 2000, and then INTEGRAL (2002), and SWIFT (2004). SWIFT in particular is a formidable tool for detecting GRB outbursts and analyzing the decay of the glow. Its primary objective is to use its BAT (Burst Alert Telescope) to discover a new GRB and relay its position (typically with a positional accuracy of between 1 and 4 arcminutes) to the ground while simultaneously slewing itself to bring the GRB within the field of its X-ray (XRT) and ultraviolet-optical (UVOT) telescopes.

At the time of writing SWIFT had detected almost 200 GRBs, and nearly all have been followed up with its onboard narrow-field instruments. A huge amount has been learned about GRBs since the start of the century, and some have suggested that the amateur's role in GRB detection is diminishing. However, history has shown that science always progresses fastest when as much data, from as many sources as possible, is collated. In addition, satellites have a limited lifetime, and professionals greatly value accurate amateur observations, so the amateur still has a role to play. Above and beyond these points there is still tremendous satisfaction to be had from imaging the dying glow of a colossal explosion at the edge of the observable universe from your own backyard.

GRBs from the SWIFT Era

GRB 050509B

R.A. 12h 36m 13.67s Dec. +28° 58′ 57.0″
Peak magnitude: N/A
Magnitude 1 day later: N/A

On May 9, 2005, a short (sub 2-second) GRB was detected that, for the first time, enabled the pinpoint precision in space of this kind of short-duration burst. The Swift X-ray telescope was used to capture the fading X-rays from the event. The GRB was tracked down to an elliptical galaxy in the cluster of galaxies cataloged as ZwCl 1234.0+02916, although some suggested that maybe the galaxies were gravitationally lensing the GRB, and it lay in the background, behind the galaxies. No optical transient was detected for this GRB, and there was no evidence of any supernova being involved. This considerably strengthened the argument that short-duration GRBs were caused by a different mechanism to the long-duration supernova-based GRBs. See the earlier short-duration GRB section for more details. Assuming the source was located within the elliptical galaxy its redshift of $z = 0.226$ places the GRB roughly 2.7 billion light-years from Earth.

GRB 050724

R.A. 16h 24m 44.40s Dec. –27° 32′ 27.9″
Peak magnitude: N/A
Magnitude 1 day later: ∼21

The July 24 2005 GRB is still one of the most intensively studied short-duration GRBs to date. The Swift X-ray telescope, NASA's Chandra telescope, and the giant Keck telescope in Hawaii all focused on the spot where the GRB had erupted. Although the GRB itself was of short duration the X-ray afterglow had a long, flaring characteristic. Once again an old elliptical host galaxy was tracked down as the most likely home of the object, the sort of galaxy that had, perhaps, more than the average share of black holes and neutron stars. Professional astronomers have speculated that a black hole swallowing a neutron star was a possible cause of this GRB. The host galaxy's redshift of $z = 0.258$ implies a distance from Earth of roughly 3 billion light-years.

GRB 050904

R.A. 00h 54m 50.8s Dec. +14° 05′ 10.0″
Peak magnitude: ∼19 in infrared
Magnitude 1 day later: ∼22 in infrared

At the time of writing the GRB detected at 01:51:44 UT on September 4, 2005, is still the most remote GRB yet detected. As with the October GRB 021004, discovered 3 years earlier, it was located in Pisces, but unlike that GRB its light was so seriously redshifted there was no hope of amateur astronomers imaging the optical afterglow. The burst duration was measured as 225 seconds, far longer than the typical 10 seconds of a long GRB. However, there is a $(1 + z)$ relationship between observer and source time durations caused by the time dilation effect, and this GRB had a colossal redshift of $z = 6.3$! This means that the GRBs light has traveled for 13 billion years to get to us, and it was emitted when the universe was far less than a billion years old. In addition, the actual burst duration was probably only about 30 seconds. Once again, the progenitor was assumed to be a massive supernova collapsing into a black hole.

GRB 060218 = SN 2006aj

R.A. 03h 21m 39.80s Dec. +16° 52′ 06.0″
Peak magnitude: SN 2006aj peaked in mid/late Feb 2006 at approx mag 17.4
Magnitude 1 day later: ~17.6

GRB 060218 occurred in the constellation of Aries, not far from the border with Taurus. The GRB had some unusual properties, including a very long burst, and its near simultaneous appearance with the supernova 2006aj indicated this was, once again, the result of a giant star collapsing into a black hole as a Type Ib/Ic supernova. The 17th-magnitude supernova was too faint for amateur visual detection, but a dozen amateurs worldwide imaged the object in February and March 2006 until it slid toward 18th magnitude and fainter. For a collapsar hypernova in a galaxy as close (a relative term!) as 440 million light-years, a peak magnitude of 17.4 was not exceptional. Nevertheless, it was yet another link between long-duration GRBs and supernovae.

GRB 060505

R.A. 22h 07m 04.00s Dec. –27° 49′ 08.0″
Peak magnitude: 22?
Magnitude 1 day later: ~22

GRB 060505 presented astronomers with another puzzle. Here was a long- (4-second) duration GRB associated with a galaxy that was not at the edge of the observable universe ($z = 0.089$, distance = 1.2 billion light-years), and yet there was no evidence for any associated supernova, even when examined with the Hubble Space Telescope. The faint afterglow was traced to a prominent star-forming region in the Sc-type galaxy with the rather technical name 2dFGRS S173Z112. This is a very similar site to that in which other long-duration GRBs have been found, but with a visible supernova. Astronomers were faced with two options. Either the physical mechanism was the same as that for short-duration GRBs (i.e., a double neutron star or neutron star/black hole merger), but for some reason the resultant GRB had a very long duration, or, it was a standard collapsar but with no visible supernova. The latter explanation could be the case if the star collapsed so quickly into a black hole that there was no time for a traditional luminous outburst. Astronomers still disagree over this one, but the evidence perhaps favors the collapsar/black hole theory, as most of the evidence, except the lack of an optical supernova, is the same as for the other collapsar/long GRB examples.

How to Observe GRBs

In the final years of the twentieth century (which was not all that long ago) detecting the optical counterpart of a fading GRB was an event that carried a lot of kudos, in both the amateur and professional communities. Amateurs were well placed, with their CCD equipment, to react quickly and image objects well below 20th magnitude. In addition there were, initially, no X-ray/gamma-ray satellites that could pinpoint a high-energy transient to better than 10 arcminute precision.

Thus, detecting a rapidly fading optical counterpart was easier if you had an amateur telescope with a wide field of view.

Amateurs had other advantages as well. They were scattered around the globe at many latitudes and longitudes, so not all of them could be clouded out at the same time, and some of those with clear skies would always be on the night-time side of Earth experiencing darkness and with the GRB field above their horizon. In addition, in the early days, before the importance of GRBs was fully appreciated, it could be quite difficult to interrupt an observing run on a world class telescope, which was already fully oversubscribed by professional astronomers with their own agendas.

Things have changed considerably as of 2008, but that does not mean the amateur no longer has a part to play. The latest generation of high-energy satellites can pinpoint GRBs quickly and to a high precision, well within the small field of view of even a 10-m telescope such as Keck I and II. In addition, the speed with which ground-based astronomers are alerted to a GRB is very rapid now, and professional astronomers at many institutions are well aware that a bright, well-placed GRB overrides all other observing programs. The modest (by professional standards) 2-m Faulkes telescopes can respond very quickly to GRB alerts and have far more light grasp than even the largest amateur telescopes.

Still, the speed with which amateurs are alerted is rapid, too, and although the field of view issue is no longer a consideration, amateurs still have advantages where the optical transients of some GRBs are concerned. With the exception of instruments such as Hubble and Spitzer (and they cannot be slewed instantly), most telescopes capable of imaging in the optical band are ground-based and can be clouded out, or in local daylight, when a GRB goes off. In addition, a far southern hemisphere facility cannot image the far northern hemisphere of the sky and vice versa.

Where a bright supernova is associated with a GRB the amateur can play a vital role monitoring the decline of the object over weeks and months, a time allocation few professional observatories can justify. But perhaps the greatest challenge of all is the visual detection of the optical counterpart of a GRB fading. The Finnish amateurs Tuukkanen and Henriksson witnessed this with GRB 030329A (described later), and a few more amateurs have witnessed the more steady light of an associated supernova from a GRB, but actually witnessing a bright GRB optical transient fading (and maybe flickering) with your own eye must be a goal that many amateurs would strive for.

To be fully prepared for the next GRB event you need to subscribe to some form of alert system. All astronomers (that I know) are on e-mail and most have cell (mobile) phones, with a few having electronic pagers, too. By giving your contact details to an organization issuing alerts, such as the AAVSO or, in Britain, *The Astronomer* magazine, you can sign up to the most convenient system.

To subscribe or unsubscribe to the AAVSO High Energy Digests visit www.aavso.org/mailman/listinfo/aavso-hen or, via email, send a message with subject or body 'help' to aavso-hen-request@aavso.org.

The AAVSO also has a high energy webpage at www.aavso.org/observing/programs/hen/ and a webpage regarding cell phone and pager alerts at www.aavso.org/observing/programs/hen/filterdatabase.shtml

In Britain, if you are a subscriber of *The Astronomer* (or, sometimes, the BAA), e-mail alerts can be issued for the most significant astronomical events. The relevant web pages are:

The Astronomer: www.theastronomer.org

British Astronomical Association: www.britastro.org

Sky & Telescope also has an astro alert service that handles a range of unusual events in the night sky:

http://www.skyandtelescope.com/resources/proamcollab/AstroAlert.html

Of course, when you slew to the region of a new GRB the CCD field will be totally unfamiliar, just a sea of stars, anyone of which might be the optical component of the GRB. If you are lucky someone might produce a finder chart and put a link to it in the AAVSO high energy e-mail alerts. However, more likely is that you will need to visit the Space Telescope Science Institute's (STScI) Digitized Sky Survey (DSS) pages at http://archive.stsci.edu/cgi-bin/dss form and download a chart of the field, to look for a new faint object in the field of view.

Notable Amateur Successes to Date

Amateur astronomers, with their new and powerful backyard equipment, were quick to latch onto the opportunities offered by their CCD detectors and Go-To telescopes. Many could image objects below magnitude 20 and, with fields of view typically 10–20 arcminutes across, the preliminary satellite positions for GRBs were almost always going to fall on their detectors. In addition, amateur observatories can be found all over the planet, and so if a GRB occurred in the northern or southern hemisphere, or when it was day or night in the United States, it made no difference; somewhere an amateur observatory would be poised to spring into action.

Arguably the first serious amateur involvement with detecting GRBs, or at least their associated supernovae, occurred in April 1998 with GRB 980425 and its associated supernova 1998bw. Amateur astronomer and supernova discoverer Berto Monard, based at his Bronberg observatory near Pretoria, made visual observations of that object as did a number of other amateurs. The Reynolds Amateur Photometry Team (RAPT) in Canberra, Australia, also monitored the object. However, it was the associated long-duration supernova they were observing, rather than the GRB afterglow itself, which, as we have seen, typically fades below magnitude 20 within a day (or less) of the initial outburst.

Warren Offutt and GRB 990123

R.A. 15h 25m 30.10s Decl. +44° 46′ 00.0″
Peak magnitude: 9!
Magnitude 1 day later: 20

We have already noted, in the earlier part of this chapter, that GRB 990123 briefly reached a staggering magnitude 9 only seconds after it was detected and then imaged by the ROTSE lenses. There is general agreement that the first amateur to actually image a fading GRB afterglow, namely this GRB, was Warren Offutt of Cloudcroft, New Mexico, in the United States. Offutt was already well known in the amateur

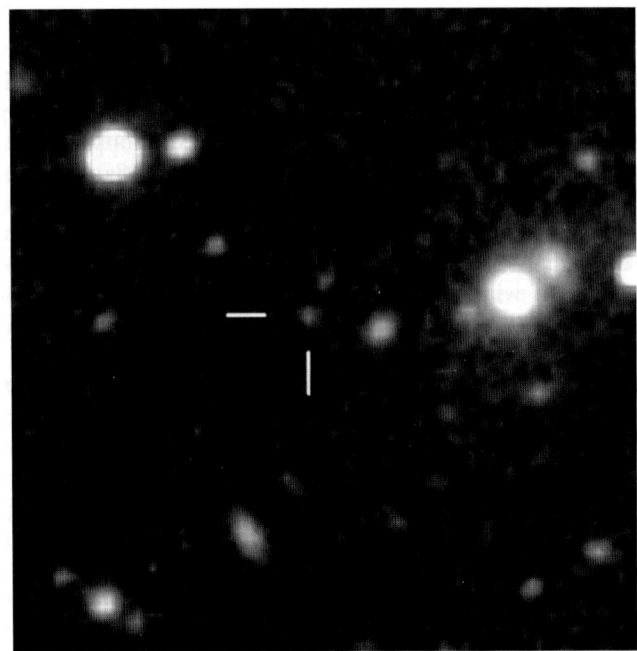

Figure 6.5. Arto Oksanen's image of the fading afterglow of GRB 000926. Image by kind permission of Arto Oksanen.

GRB 010222

R.A. 14h 52m 12.00s Dec. +43° 01′ 06.0″
Peak magnitude: 18
Magnitude 1 day later: below 20

On February 22, 2001, at 07:23:30 UT one of the brightest GRBs ever observed by the BeppoSAX satellite was detected simultaneously by its GRBM and WFC1 instruments; the GRB was in the northern part of the constellation Bootes. Slightly more than 4 hours later professional astronomers at the Harvard-Smithsonian Center for Astrophysics announced they had imaged the fading 18th-mag afterglow with the F. L. Whipple Observatory's 1.5-m Tillinghast telescope. At the same time, amateur astronomer Gary Billings of Calgary, Canada, imaged the fading GRB, too, with a Celestron 14 and AP7b CCD. Some 12 hours after the initial GRB the Finnish amateurs of the Nyrola observatory were successful again. Amateur astronomers Oksanen, Moilanen, Hyvonen, Pasanen, and Tikkanen observed the optical transient with their 0.4-m LX200 telescope and ST7E CCD camera recording its fade from 19th to 20th magnitude.

GRB 021004

R.A. 0h 26m 54.69s Dec. +18° 55′ 41.3″
Peak magnitude: ~15
Magnitude 1 day later: 19

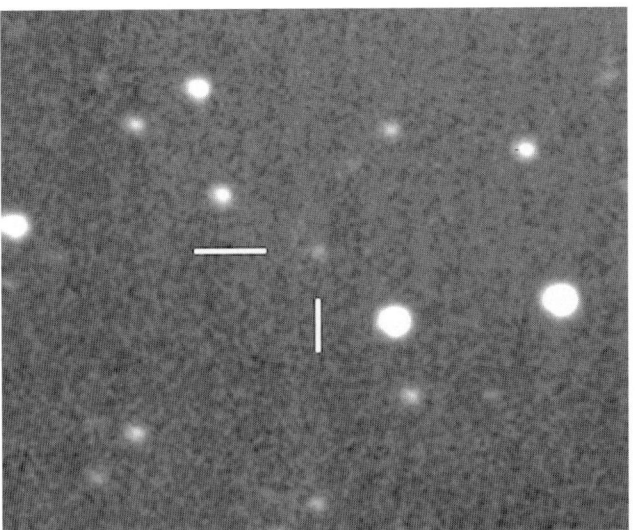

Figure 6.6. An image by the author of GRB 021004, which occurred on October 4, 2002, and was detected by the High Energy Transient Explorer 2 (HETE 2) satellite in the constellation of Pisces, just to the East of the famous 'Square of Pegasus.' 0.3-m f/3.3 Schmidt–Cassegrain and SBIG ST7 CCD. 20 × 100 second exposures.

On October 4, 2002, at 12:06:13.57 UT the High Energy Transient Explorer 2 (HETE 2) satellite detected a GRB in the constellation of Pisces, just to the east of the famous 'Square of Pegasus.' For many British amateur astronomers, including this author, this would prove to be the first realistic chance to image the fading optical afterglow of a GRB (see Figure 6.6). The other amateurs were Nick James, Mark Armstrong, Tom Boles, David Strange, Peter Birtwhistle, and Eddie Guscott. A redshift of 2.3 was measured for the fading optical transient, and this translates to a distance of about 11 billion light-years, assuming a Hubble constant of 70. Once again, a massive Wolf–Rayet progenitor star ending its life as a collapsar (supernova collapses to a black hole) was thought to be the likely cause of this GRB. More than 50 hours after the initial GRB British amateur Tom Boles was still able to image the object with a 35-cm Celestron 14 telescope at almost 21st magnitude.

GRB 030329A = SN 2003dh

R.A. 10h 44m 50.03s Dec. +21° 31′ 18.15″
Peak magnitude: 12
Magnitude 1 day later: 16

Less than 6 months after GRB 021004 European and British amateurs had another opportunity to image a fading GRB, one that was still surprisingly bright as darkness fell on March 29. The burst was detected by the HETE 2 satellite at 11:37:14.67 UT on March 29, 2003, in Leo, roughly a degree south of the constellation border with Leo Minor.

Two hours after the outburst, observers at Kyoto Observatory, near Tokyo, Japan, spotted a magnitude 12.4 fading afterglow. Less than 9 hours after the initial burst two amateur astronomers in Finland, as mentioned earlier, observed the optical counterpart visually, using modest aperture telescopes. They were members of the Finnish Deep Sky section of the Astronomical Association 'Ursa.' M. Tuukkanen observed it in Pornainen, Finland, with a 0.63-m Newtonian telescope for about 1 hour starting at 19:30 UT. He described it as a faint starlike object easily seen with direct vision. Tuukkanen did not see any flickering or a distinct color. R. Henriksson was observing in Orivesi, Finland, with a 0.30-m Newtonian telescope at ×200 magnification (at 20:05 UT). Henriksson reported the object as stellar and faint, visible only with averted vision. Both Finnish observers estimated the optical transient magnitude as 14.3 by using the 14.2-magnitude star GSC 1434:322, just north of the GRB, as a reference. This information was originally reported by their countryman and fellow GRB hunter Arto Oksanen of the Nyrola Observatory, on GCN 2010. Much of the rest of Europe was clouded out that night, but the brightness of GRB 030329 and its relatively slow decline meant that many other amateurs, including this author, were able to image it on the next night, March 30, when it was still 16th magnitude. As well as myself, Peter Birtwhistle, Tom Boles, David Strange, John Fletcher, and Nick James bagged GRB 030329 on March 30 and 31 in 2003.

On March 31 the GRB optical counterpart was, remarkably, still magnitude 16 (see Figure 6.7) and had only faded to 17 by April 1. By the night of April 7/8 the object was still above magnitude 19 (just). As late as April 23 the British observer Peter Birtwhistle imaged the optical counterpart of GRB 030329 at magnitude 20.4, 25 days after the GRB went off. In fact, it turned out that the GRB could be unambiguously associated with a Type Ic supernova, SN 2003dh, whose signature

Figure 6.7. An image of GRB 030329/SN 2003dh. 4 × 100 seconds with a 0.3-m SCT at f/6.3 and SBIG ST9XE CCD taken by the author on March 31, 2003. The field is 18 arcminutes wide, with north at the top. The GRB was unambiguously associated with the Type Ic supernova, whose signature was quickly identified in the spectra of the fading afterglow.

was quickly identified in the spectra of the fading afterglow. Indeed, from the point of view of optical photometry, the object that was being imaged worldwide by amateur astronomers immediately after the GRB trigger was partly the GRB and partly the supernova.

Ever since the bright, 14th-magnitude, Type Ic supernova 1998bw in the galaxy ESO 184-G82 had been associated with GRB 980425, the case for some GRBs being associated with giant collapsar supernovae had been strengthened. However, some doubts had still remained simply because there had been many cases of Type Ic supernovae where a GRB had not been detected, even in, relatively speaking, nearby galaxies, such as M74/NGC 628 (SN 2002ap) or even SN 1997ef in UGC 4107. It was becoming clear that in the cases where collapsar-type supernovae had not produced a GRB the jets were (probably) simply badly aligned with Earth. Someone else in the cosmos might be able to see the GRB, but not us. Because of the prolonged brightness of GRB 030329/SN 2003dh and the previous example of GRB 980425/SN 1998bw, astronomers were now far more confident that long GRBs and collapsar supernovae were linked, and the GRB980425/SN 1998bw event was not an isolated case. A redshift of $z = 0.1685$ was determined for the GRB030329 event, corresponding to a distance of more than 2 billion light-years.

GRB 071010B

R.A. 10h 02m 09.26s Dec. +45° 43′ 50.3″
Peak magnitude: at least 17
Magnitude 1 day later: unknown

Even as this book was being written, despite the Swift satellite now regularly detecting GRBs, and precisely slewing to point at GRBs, amateurs can still occasionally make a significant contribution. On October 10, 2007, at 20:45:47 UT, the Swift Burst Alert Telescope (BAT) triggered and located GRB 071010B. The BAT onboard calculated location was given as R.A. = 10h 02m 07s, Dec. = +45° 43′ 53″ with an uncertainty of 3 arcminutes.

The BAT light-curve showed a single pulse whose duration was about 20 seconds, peaking in intensity roughly 2 seconds after the trigger. At the time of this GRB Swift was in the process of returning to normal operations for technical reasons, and so the automatic slewing to GRBs was disabled outside of business hours in the United States. There were therefore no X-Ray or UV observations of this burst from the onboard instruments. This opportunity was not wasted by, yet again, the Finnish amateur Arto Oksanen, who secured images of the optical afterglow at 21:11:05 UT, only 25 minutes after the Swift trigger! His observations, from Hankasalmi Observatory in Finland, showed the transient fading from 17th to 18th magnitude and provided a precise position of R.A. 10h 02m 9s.26, Dec +45° 43′ 50.3″ – yet another success for amateur astronomy.

Final Thoughts

Arto Oksanen's continuing success in this field, and the rarity of visual observations of fading GRB counterparts, means that, even in the SWIFT era, swinging a telescope rapidly on the field of a fading GRB is still a worthwhile pursuit for the amateur astronomer. Remember GRB 990123? This is still the only GRB to have been detected seconds after the burst with a magnitude of 9. What a coup it would be for an amateur astronomer to bag a second event of this brightness, or maybe one even brighter! Remember, these events fade rapidly, but amateurs can respond rapidly. A serious backyard observer, with a telescope ready to slew, will, one day, get an e-mail while he or she is imaging another object, slew to the field, and bag another fading GRB of 9th magnitude. Maybe someone who knows the sky intimately will even be able to catch it fading in large binoculars?

Chapter 7

How to Do Visual and CCD Photometry

Visual Photometry

Although we are now in the twenty-first century, when many leading amateurs use CCD cameras for photometric measurements, there are still plenty of visual observers who make magnitude estimates using their own eyes and good judgement, with no technical wizardry. Using big Dobsonians, typically of 0.4-m or 0.45-m aperture, or Schmidt–Cassegrains of 0.35-m aperture, they can make estimates down to 16th magnitude. Many Dobsonian users can locate their targets more efficiently than using a motorized Go To system, as they can haul their telescopes around the sky much more quickly than a few degrees per second and without the fear of any gearbox failures (common with heavy out-of-balance fork-mounted Schmidt–Cassegrains). The visual approach also puts the observer in touch with reality in a way that the CCD observer never can be: photons from the outbursting dwarf nova are actually hitting the retina!

Of course, it takes years to gain sufficient experience so that you can, without reference to finder charts, starhop through dozens of fields and make a magnitude estimate in seconds, let alone one of a 15th- or 16th-magnitude quarry! In addition, the large apertures necessary to see down to such magnitudes necessitate large telescope tubes, unless a compact Schmidt–Cassegrain is used. An f/4.5, 0.4-m Dobsonian will have a tube length similar to its focal length of 1.8 m, necessitating a ladder to reach the highest eyepiece altitudes and some neck-twisting observing positions. Observer discomfort is not good for seeing down to those very faint targets, and if you are a beginner, craning your neck to get to the eyepiece while holding a chart in one hand and a red torch in the other is no fun at all.

Alt-azimuth-mounted Go-To Schmidt–Cassegrains offer an ergonomic alternative for the visual observer, with the big advantage that the eyepiece moves little wherever the telescope is pointed. However, if you use the supplied star diagonal you may need to re-familiarize yourself with all the star fields, as the view through the eyepiece will be mirror-flipped. This latter issue can be solved by purchasing a 45-degree erecting Amici prism, which will turn the view into a terrestrial one (the right way up) but will not mirror-flip the view. A thoughtful arrangement of finder telescopes on any large instrument can make locating an object in the main field a lot simpler, too. As a general rule visual observing is best when two criteria are satisfied: namely (1) the telescope should have as large an aperture as possible, and (2) the telescope should be as user-friendly as possible. Unfortunately these two criteria often prove to be mutually exclusive!

A few observers do seem to have truly exceptional night vision, but, in the main, this is very rare, and if you cannot see the faint objects that others can it is

M. Mobberley, *Cataclysmic Cosmic Events and How to Observe Them*,
DOI: 10.1007/978-0-387-79946-9_7, © Springer Science+Business Media, LLC 2009

probably due to your local light pollution, impatience, a poor night, or inadequate dark adaption. Choosing the right magnification, collimating your telescope, re-aluminizing a Newtonian's mirrors, and, specifically, mastering the use of 'averted vision' are all crucial.

The Retina

The eye has two types of detectors within the retina. These are called rods and cones. The rods are the low-light detectors, whereas the cones allow full-color high-resolution eyesight. A small central portion of the retina is packed with cones, which you are using to read this sentence. Your brain creates the illusion that the whole book page is sharp, but in fact you are only seeing a few letters at a time at high resolution and in full color; it is your eye muscles that are swiftly zipping everywhere and creating the illusion.

The electrochemical signals from the retina's detectors travel via cells known as ganglion cells on their way to the brain. In the high-resolution, full-color retina center (the fovea), one ganglion cell interfaces to one cone. But, as you go further out and low-light rods dominate, there may be 100 rod detectors passing their electrochemical signal into just one ganglion cell. It is a case of paralleling up to improve the signal-to-noise ratio.

Not surprisingly, with so many detectors being ganged together, resolution suffers badly. Although the foveal cones can resolve 1/60th of a degree (one arcminute), the ganged rod system well away from the center might only resolve 20 arcminutes; this is not much finer than the size of the Moon seen with the naked eye. There is an optimum, ultrasensitive, rod-packed region of the retina that is roughly 8–16 degrees away from the eye's center; 12 degrees is a good average value for the best part. This means that, to get the most sensitivity out of your retina, you have to look to one side of the faint astronomical object you are trying to see; at first this will seem incredibly difficult, but it will improve with practice. This 12-degree (or so) offset should be arranged so that you appear to place the object nearer to your nose! The reason for this strange requirement is that the eye has a blind spot where the optic nerve leaves the retina, and this blind spot is on the other side, away from the nose. (Actually, it isn't! The eye's lens turns everything upside down, but we are looking at how it feels here and not how it actually is!) Roughly speaking the eye is 4 astronomical magnitudes (40 times) more sensitive at this crucial point than in the center. So if you can hold, say, a 10th-magnitude star steady in the visual center of a 30-cm reflector's telescope field you can hold a 14th-magnitude star steady on the rods 12 degrees off center.

Dark Adaption

Of course, when you first go outside and look through the eyepiece you will probably not see anything. This is because you are not 'dark adapted.' When the human eye is plunged into darkness, two things happen. First, the pupil dilates (expands) to its maximum diameter. In young people this may be 7 mm or so, but for astronomers in their eighties it may only be a few millimeters across. This is not a big problem at the telescope, as a higher magnification produces a narrower beam of light that can pass through a small pupil. A tighter beam of light will also be less affected by astigmatism in the older observer's eye.

The second development in darkness is that the amount of the chemical rhodopsin in the retina increases dramatically, by many thousand-fold. So the combined effects of rhodopsin and using averted vision amounts to more than 100,000 times more sensitivity than your central vision had in a fully illuminated room before you stepped outdoors. Dark adaption – waiting for the rhodopsin to do its job – cannot be rushed. You need to wait 40 minutes or more to feel the full effect. So, if you are planning to observe a number of faint objects, save the faintest ones till the very last!

Making a visual brightness estimate (known as a magnitude estimate) of a faint dwarf nova or a supernova is, in theory, no different from estimating the brightness of any other variable star. However, with many dwarf novae and supernovae being exceptionally faint, even in outburst, estimating their brightness is a major challenge in all but the largest amateur telescopes. However, despite these problems, dedicated visual observers, like Gary Poyner, can make thousands of magnitude estimates of faint cataclysmic variables (CVs) and supernovae throughout the course of a year.

The eye is a remarkable detector, but it does take a while to reach full sensitivity. For many years I used a massive 49-cm (19.3″) Newtonian at my former residence near Chelmsford in England. The telescope was right outside my living room door, so I was in action just a few minutes after leaving the house. However, I was far from being dark-adapted by that time. Somewhat surprisingly I found that finder charts showing stars down to a mere magnitude 11 were the best ones to use to locate the right field. After half an hour and, at very high powers, I could detect stars of magnitude 15, but for that initial location of the target a wide-field chart to magnitude 11 was ideal. Spending a day outdoors in the Sun is not a good idea if you are a faint variable star observer, unless you wear wrap-around sunglasses all the time. The eye takes 24 hours to recover from such an onslaught, and you will lose 0.7 or 0.8 magnitudes of your sensitivity the following night however long you try to dark adapt. The usual 30- or 40-minute dark-adaption routine will not work.

It might instinctively be expected that the 'exposure time' of the eye was somewhere in the order of a fifth of a second, not dissimilar to the human reaction time. Although this is true for daylight observations, in darkness things are rather different. At the faintest levels it pays to stare at an object such that the flow of photons hitting the rods is sufficiently high enough for several seconds to trigger a definite 'hit' in the brain. In practice, at a typical observing site, with background light pollution, this level is usually reached when several hundred photons per second are hitting a group of rods.

This staring action should not be interpreted as an exposure time as such. The eye is not a digital device like a CCD, but it certainly pays to stare at a faint object to see if it emerges. What the trained and dark-adapted eye sees, after studying an object patiently, is an impression not dissimilar to what a low-resolution CCD would capture with an exposure of a few seconds. Of course, at these faint levels we can hardly ever make an accurate magnitude estimate of a star; all we can do is say we have seen it and maybe estimate its magnitude in relation to a slightly brighter comparison star.

Let us return to something mentioned earlier too. In elderly observers the eye's pupil does not dilate as much as the 7 mm that the young eye is capable of. This means that a loss of light results at very low magnifications. The bundle of light rays from the eyepiece will have a diameter equal to the telescope aperture divided by the magnification. If this is larger than 7 mm, then light will be lost however young you are. Older observers will find that 5 mm is a more sensible figure to adopt. In other words, with a 300-mm aperture telescope, magnifications of $\times 60$ or more are highly recommended. Much higher magnifications are typically used for studying faint stars, and the diameter of the pupil is really only a consideration at low powers when you are probably just locating the field anyway.

Many newcomers to astronomy think that to see fainter objects you need a lower magnification. This misconception may come from the fact that on bright objects such as the Moon, as you whack in a higher eyepiece the view just gets increasingly faint. However, different rules apply for the faintest objects. For a start, contrast becomes just as important as brightness. Only the luckiest amateur astronomers enjoy a really dark sky, and for many town dwellers the night sky has a sickly orange glow, from the thousands of nearby streetlights. As you increase the telescope's magnification, the background skyglow becomes much dimmer, and point sources such as faint stars (typically spanning a few arcseconds in diameter due to atmospheric turbulence) start to cover maybe just a few more rods, convincing the brain that something is really there. Both of these factors enable the faintest stars to be seen more easily. Also, as the field of view becomes narrower, there is less chance of any really bright stars encroaching into the eyepiece field and dazzling the observer.

At this point we should mention something that is puzzling. Most knowledgeable amateur astronomers will tell you that a telescope's focal ratio has no bearing on how bright the background sky appears in a light polluted area. The sky background will simply get dimmer and blacker as you whack up the magnification. If an f/6 and an f/4 Newtonian of the same aperture are used at the same magnification, the sky background brightness should look the same. Obviously, to match the magnifications, you would have to use two different eyepieces, that is, a 4-mm eyepiece with the f/4 instrument, and a 6-mm eyepiece with the f/6 instrument. However, if you ask really experienced visual observers who live in highly light-polluted areas what their opinion is, they will tell you that the slower f-ratio telescope (f/6 in this example) always makes the sky darker, even at the same magnification. So this has to be an issue more related to scattered light in the tube than a straightforward optical issue.

The f-ratio of a telescope should have no bearing on the background sky brightness for a specific magnification. The same number of photons will hit the same number of rods and cones when aperture and magnification are identical, unless there is something different in the design of the two telescopes or the two eyepieces. Nevertheless, this 'slower is darker' rule seems to apply and has been supported by an observer who has made over 200,000 magnitude estimates down to (almost) magnitude 17, so we have to take it seriously. The performance of eyepieces on telescopes with fast optics may have a bearing on this issue, too.

Just how little light can the human eye detect? Well, for typical observers in good conditions the standard formula that is most often used is $2 + 5 \log_{10}D$, where D is the telescope diameter in millimeters. However, nothing in life is simple, and experienced observers can see well below the magnitudes predicted by the standard formula. Indeed, some can see stars to almost magnitude 8 at the zenith simply using the naked eye at a truly dark site. Unfortunately human beings are often impatient and intolerant of things that cannot be stated precisely in black and white. People like a straight answer to a straight question and not 'Well, it all depends on' types of answers.

The plain fact is that every observer has a different degree of experience, a different amount of light pollution, a different set of optics, and a different pair of eyes. So the answer to 'How faint can I see?' has to be 'Go outside at night and find out!' One authority on human night vision, Roger N. Clark, has pointed out (*Sky & Telescope*, April 1994, pp. 106–108) that a study conducted by H.R. Blackwell, in 1946, likened the eye's detection ability, at its limit, to a probability curve. To quote specific examples, Clark proposed that with a 400-mm-aperture telescope an observer working at the limit, in excellent conditions, could just glimpse a magnitude-15.7 star 98 percent of the time, a magnitude-16.7 star 50 percent of the time, and a magnitude-17.7 star 10 percent of the time. The 50 percent probability level corresponds to the formula $3.7 + 5 \log_{10}D$ and the row labeled 'Clark' shown in Table 7.1.

In fact, Clark's 50 percent probability formula corresponds very well to the sorts of magnitudes reached by the world's most experienced visual CV observer, Gary Poyner, who has glimpsed stars well into the high-magnitude 16 s with a 35-cm Schmidt–Cassegrain and with 40-cm or 45-cm Dobsonians. The other rows in Table 7.1 correspond to a formula proposed by Bradley E. Schaefer (NASA-Goddard Space Flight Center) in 1989 after an extensive survey, and a formula proposed by the author and amateur astronomer Gerald North after studying Schaefer's report. Of course, there are always the really freaky results that amateurs tend to remember, for example, Stephen J O'Meara's detection of comet Halley, visually, at magnitude 19.6, with a 60-cm telescope at Mauna Kea in Hawaii in January 1985, while breathing oxygen! He also identified a magnitude-20.4 field star at the same time. Although this sounds impossible, he had spent '1–2 hours' staring at the field waiting for a definite 'glimpse.' The feat was verified by O'Meara describing field stars with no prior knowledge of their position. This sort of detection feat would correspond to about 1 or 2 percent on the Clark 'probability level.'

In practice, experienced observers can often reach 2 magnitudes fainter than the standard formula for stars at high altitude and on the clearest nights, even from urban locations. In fact, rods are so sensitive that they can actually detect single photons. In 1942 Selig Hecht proposed this because light flashes so dim that only 1 percent of rods likely to absorb a photon were detectable by observers in experiments. Experiments by Schneeweis and Schnapf in 1995, using monkey

Table 7.1. Various visual magnitude limits, for telescope apertures in millimeters, predicted by formulae proposed by astronomers in recent years

Formula	100 mm	150 mm	200 mm	250 mm	300 mm	350 mm	400 mm	450 mm	500 mm
Standard	12.0	12.9	13.5	14.0	14.4	14.7	15.0	15.3	15.5
North	13.3	14.1	14.6	15.1	15.4	15.7	15.9	16.2	16.4
Schaefer	13.4	14.3	14.9	15.4	15.8	16.1	16.4	16.7	16.9
Clark	13.7	14.6	15.2	15.7	16.1	16.4	16.7	17.0	17.2

The standard formula has always been $2.0 + 5 \log_{10}D$ (where D is the telescope diameter in millimeters) despite the fact that many observers can see fainter. Gerald North's formula ($4.5 + 4.4 \log_{10}D$) was first suggested by him in the *Journal of the British Astronomical Association* in 1997 [**107**(2), 82 (1997)]. It fits Schaefer's results rather better for a wide range of apertures than do other formulae and, although more optimistic than the standard model, becomes slightly more pessimistic as very large apertures are used. The third row uses Bradley Schaefer's own prediction ($3.4 + 5 \log_{10}D$), while Clark's ($3.7 + 5 \log_{10}D$) is more optimistic and ties in with his 50 percent probability criteria rods, confirmed that single photons could trigger a response. The arrival of a few photons per second at the eye is the sort of rate that an observer using a 400-mm telescope might receive from a 20th-magnitude star, making O'Meara's famous Halley observation seem more feasible! Fortunately, the human retina has no electronic readout noise and does not need to be cooled.

Truly, the human eye is a quite remarkable detector. Even in the modern era of CCDs and webcams its versatility is extraordinary. It can cope with illumination levels from bright sunlight to starlight, spanning 100 million-fold in intensity, and even survey almost the whole night sky for meteors, following them instantly as they whiz across the sky. Amazing!

Estimating Magnitudes Visually

To estimate the brightness of any variable star, however bright, it is necessary to compare it with other nearby stars of a known magnitude. Obviously, the comparison stars themselves must not vary in brightness, or all hope for a useful measurement is lost. Fortunately, for established dwarf novae, a reliable photometric sequence is usually available, and organizations such as the British Astronomical Association (BAA), TA (*The Astronomer* magazine), and the American Association of Variable Star Observers (AAVSO) can supply charts for dozens of CVs. The relevant websites are listed in the Resources section of this book. However, some dwarf novae are very faint, and new ones are being discovered regularly. Sometimes amateur observers who are good at filtered CCD photometry are called on to produce new charts, and these often need revising in the early days of studying a new object.

It can be tempting to simply use a planetarium software package to produce a homemade chart that you can use at the telescope. Although the software may be perfect for making your own customized finder charts, it is very questionable for making the magnitude estimate. Planetarium packages often have highly erroneous star magnitudes (the Hubble Guide Star Catalog is full of them), and they do not tell you if a comparison star may be highly colored or slightly variable. A photometrically verified magnitude sequence should always be used where possible, and the chart title/origin (i.e., AAVSO) and issue date/status should always be recorded. Photometric sequences are occasionally revised at a later date, so a complete record of which stars and which chart you used for your magnitude reduction is essential.

Charts for variable star observing come in a variety of formats, depending on how large your telescope is and what magnification you are using. For Schmidt–Cassegrain users they are invariably available in a left-right flipped format, to take account of the mirror diagonal often used with these instruments. Obviously, a variable star chart should be optimized to enable the field of view to be recognized and the star in question located.

Having a homemade star chart that precisely matches the widest eyepiece field and shows stars to the same limiting magnitude as that first glimpse is essential. When you initially look through a telescope eyepiece you are rarely perfectly dark-adapted. In addition, what you are trying to do is to identify star patterns. If you can only see the brightest stars on your chart you may easily find a bright pattern of stars that looks surprisingly like a fainter pattern on the chart. When you go to the eyepiece you must have total confidence that you know which way up is the north point and that the field of view exactly matches the chart field. For very faint stars and supernovae you may well need a couple of charts, that is, a wide-field custom chart to find the field and a narrower-field, higher-magnification chart for making the magnitude estimate.

It is surprising how few faint stars you can see if you have only been dark-adapted for a matter of minutes. If you do not have a Go To telescope, mechanical setting circles fitted to an equatorially mounted telescope can give you enormous confidence that you have the right field. Even a declination circle on its own can be a very valuable aid, especially for finding objects in twilight.

Magnitude Reporting

In practice, experienced variable star observers use one of two methods to estimate the magnitude of a variable star. These are known as the fractional method and the Pogson step method. Although magnitudes are generally quoted to one decimal place, i.e., 13.1 or 13.7, in practice visual accuracies of ±0.1 magnitude (equivalent to ±10%) are usually achieved by luck, at least where really faint stars are concerned. If an observer thinks his or her magnitude estimate is likely to be accurate to within ±0.1 magnitudes it is known as a Class 1 estimate – in other words, as good as it can be. The human eye and brain is not a photometry machine; it is really designed to cope with extremes of brightness and has a largely logarithmic response. Magnitude estimates are just that estimates to, at best, the nearest 10 percent, and usually far worse. This, though, is not a real problem, as many variable stars show considerable variations and even a measurement of ±20 percent accuracy is very useful. The precise definitions of the classes of variable star estimate, as defined by the BAA Variable Star Section (VSS) are as follows:

Class 1: Very confident of the estimate made under ideal conditions and confident of an accuracy of 0.1 magnitudes.

Class 2: Less confident than 1, maybe cloud or tiredness or stray light interfering. An accuracy of 0.2 magnitudes.

Class 3: Observation made under extremely poor conditions; variable just glimpsed a couple of times; an accuracy of 0.3 magnitudes or less.

Probably the only way to achieve a Class 1 estimate would be if the two comparison stars were 0.1 magnitudes brighter and fainter than the bright variable star being estimated. But let us now have a look at the actual estimation techniques.

Fractional Estimates

The fractional method is very easy to understand. Imagine you have a supernova of roughly magnitude 14. You have already acquired a star chart from the AAVSO or BAA, and you spot two suitable comparison stars on that chart. One is labelled E and has a magnitude of 13.4. The other is labelled F and has a magnitude of 14.2. Armed with the chart and a dim red light you approach the telescope drawtube, replacing the low-power eyepiece you used to find the field with a high-power one. After locating the supernova and the two comparison stars you turn the red light off and stare at the field, using your newly acquired 'averted vision' techniques. Imagine that the supernova appears to be a bit fainter than star E, but a lot brighter than star F. In fact, it is roughly a quarter of the way in brightness from star E to star F (which makes it about magnitude 13.6). Confident of this measurement, you turn your dim red lamp back on and make the following observation in your log book: E (1) v (3) F = 13.6.

susceptible to cold temperatures, as Arctic and Antarctic explorers who have suffered frostbite will testify! But one can overinsulate the main torso, leading to the observing session being quite a sweaty one if the observer is dressed up like a Christmas turkey. I have always preferred to have a nearby warm room to retire to for 20 minutes than to have to spend 15 minutes dressing up before and 15 more minutes dressing down after an observing session.

For visual observing, a rugged observing log book is essential, too, something that will not fall apart or blow away in a breeze and that will last for a year or so of steady observing. A variety of reliable ballpoint pens, felt tip pens, and pencils should be available, too. In cold or damp weather, 90 percent of writing implements refuse to make any mark on the paper. If a variety of pens are available they cannot all freeze or get damp! (Actually, they can, but we are trying to be optimistic here.) A solid writing surface, at waist height, should also be available, as well as a chair. Continually bending down to write on a notedpad, or kneeling on a concrete floor, does no one's back or knees any favors.

We have already mentioned the advantages of having a dim red torch (because red light dazzles the eye far less, and the night vision remains intact) and a red bicycle rear lamp works fine. However, many observers prefer a head flashlight of the sort worn by miners (with a strip of red plastic over the lamp); there is also a device made by Black & Decker, called the Snake Light. This is a flashlight attached to a long, flexible (but not floppy) hose with a battery compartment at the end. The batteries act as a nice counterweight to the flashlight end. The Snake Light hose can be wrapped around the observer's neck, with the battery end on the chest and the flashlight itself positioned between shoulder and ear. This system works well and does not seem to bang into the eyepiece end of the telescope, as a miner's lamp tends to do.

The observer needs somewhere to store star charts, too, where they will not get damp. An ideal solution to this is a loose-leaf ring binder folder with transparent page-holding pockets.

CCD Photometry

The amateur astronomy CCD revolution has brought precision photometry within the reach of the backyard amateur. Indeed, as we have already seen in Chap. 1, the Center for Backyard Astrophysics (CBA) has been exploiting backyard photometric possibilities since the 1970s, although initially this was using photomultiplier tubes.

Unlike the human eye/brain combination and its struggle to achieve better than ±20 percent accuracy with photometric measurements, the CCD can reliably achieve ±percent even in a beginner's hands and ±5 percent with just a bit more care. Indeed, in the hands of an expert much higher precision is possible. Beyond all this, though, the CCD detector has one other huge advantage: it can integrate light for many minutes, leading to a far greater signal-to-noise ratio than is possible with the human eye. In addition, images can be stored for eternity, allowing a full re-evaluation if any doubts are cast on the original magnitude estimate.

There are a number of critical factors that need addressing before a CCD image can be used for an accurate photometric measurement. We will briefly discuss these issues before dealing with them in more detail later. The first factor is one of

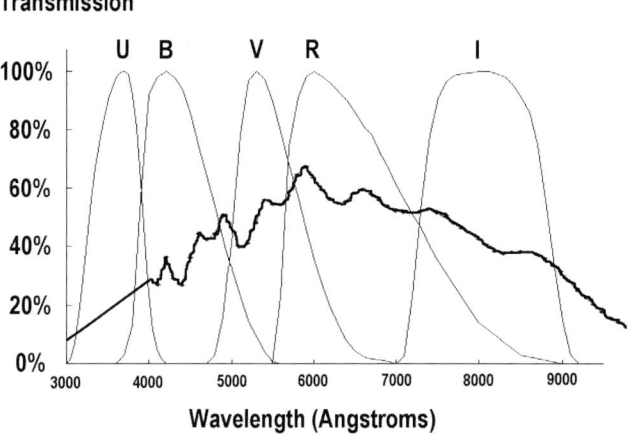

Transmission

U B   V   R   I

Figure 7.1. The transmission characteristics of the SBIG ST9XE's KAF0261E CCD chip overlaid on the UBVRI Bessell profiles. Note how the CCD sensitivity peaks in the red part of the spectrum.

color. A CCD detector has a far wider spectral range than the human eye. It can see deep into the near infrared and into the ultraviolet, too. For astronomers this is a problem, as stars can behave differently in, for example, the infrared compared to their performance in the normal human visual range.

Professional astronomers take measurements in specific wavebands called U, B, V, R, and I (Ultraviolet, Blue, Visual, Red, and Infrared), which enables them to precisely define a star's performance at each color and to calibrate their photometry accurately (see Figure 7.1). The V band corresponds to the center of the human visual range, that is, to the color green (the peak response of the eye in darkness is slightly different, but we will ignore that subtlety). At first the use of filters may seem a backward step. A CCD is so much more sensitive than the eye or film, but then we have to make it less sensitive by slapping filters in front of it. However, it is all in a good cause: the quest for scientific accuracy. Having said this, where faint dwarf nova photometry is concerned filters are frequently abandoned. The simple fact is that when 16th-magnitude objects are being monitored for subtle fluctuations, a filter, causing as much as a 2-magnitude loss in sensitivity, can push those fluctuations into the noise. In addition, for the simple detection of superhumps, or the revealing of some sort of periodicity, filters are unnecessary.

The same argument applies to the detection of eclipses or, outside the CV world, to determining the rotation rate of asteroids. When reporting unfiltered CCD observations a 'C' symbol is usually used after the magnitude estimate. V, B, and U are used for CCD measurements with Johnson V, B, or U band filters. R and I filter measurements are usually suffixed with a small j or a small c, depending on whether those filters correspond to the Johnson or Kron–Cousins photometric system. The abbreviation CR is sometimes seen and refers to an unfiltered CCD measurement that has been 'zero point adjusted' to conform to a Kron–Cousins R band filter measurement. We will delve more deeply into the world of UBVRI shortly, but first there is another issue to address.

The second critical factor relates to the linearity and usable range of a CCD. Much CCD photometry is based on comparison with a single star. Both the comparison star and the star being measured should be well above the noise floor of the image and well below the potentially nonlinear saturation point. In general CCDs are very linear devices, with the charge collected in each pixel being accurately related to the number of photons collected. However, in devices with anti-blooming gates (ABGs) appreciable nonlinearities creep in as the pixel 'well' becomes half-full, that is, half way to saturation and white-out.

ABGs are generally not found on purely scientific CCD detectors. Their sole role is to drain away excess charge, much as a storm drain copes with a surge of water. This draining activity prevents any likelihood of a very bright star 'bleeding' down the image, and so is vital for commercial applications where a pretty picture is the aim and not a scientific measurement. When an ABG first starts to work it is siphoning vital data away from your measurement of a bright star in the field, not what we want. Rather predictably, non ABG detectors are designated by the abbreviation NABG. In short, ABG CCDs should only be used for photometry when the comparison and target star are filling no more than 50 percent of the available range of the detector.

As a first step the linearity of a CCD detector can be verified by experiments using, for example, three stars for the test, that is, a faint comparison star, a mid-range target star, and a bright comparison star. This was the method I used for my very first CCD camera at the start of the 1990s. A much quicker method of verifying your CCD camera's suitability for photometry is simply to join a like-minded group of people, such as the BAA VSS (http://www.britastro.org/vss/) or the AAVSO and join a user group where ideas and tips can be freely exchanged. Someone in such a group will, almost certainly, be using the same camera as you have for photometry.

When using two comparison stars to measure an initial test target, they should, ideally, be just above and just below the target star's magnitude, as this will allow accurate calibration of the camera's linearity for a given exposure. If your camera software truly allows you to measure the digital output from the A/D converter (prior to any potential software tweaks to adjust camera performance), you can be more confident that you are aware of your system's own deficiencies. In practice the major manufacturers of scientific NABG devices ensure this, but with more 'pretty picture' application cameras, such as DSLRs, nonlinearity is a concern. It is also imperative that no image processing routines (except basic dark frame subtraction and flat-fielding) are carried out prior to the photometric analysis.

Photometry in Detail

We have just seen that color and linearity are major factors that affect the accuracy of a CCD magnitude estimate. These are not the only issues, though. Measuring the brightness of a star with a CCD full of pixels is not dissimilar to measuring the amount of rainwater in a field full of buckets. If some of the buckets (pixels) are literally overflowing (with a bright star's charge), then we have an inaccurate measurement. If there are only a few drops of rain in the bottom of the bucket the measurement will be almost impossible to make accurately (equivalent to measuring a faint star), especially if the bucket was damp to start with. Also, if some of the buckets are covered with obstructions we have even more problems.

Before an image of a variable star is exposed through the appropriate filter a few preparations need to be made. First, the star you are measuring needs to ideally fall in the mid-range of the image. For a typical 16-bit ($2^{16} = 65,536$) measurement, the maximum range will actually be 65,535 ADUs (analog to digital units) on an NABG camera. If the target star and comparison star are, say, giving readings of between 20,000 ADU and 40,000 ADU that is excellent: well above the noise and well below saturating the detector. In practice, things will never be quite that easy, and, where a target is well below the faintest star on the photometric sequence, it is quite possible that the faintest comparison star may end up registering 50,000 or more ADUs, with the target down in the murk at the 5,000-ADU level, even in a deep exposure. In other words, a supernova (for example) might be 2.5 magnitudes fainter than the comparison star, so the accuracy will suffer. Some prior knowledge of the number of ADUs produced by a given exposure through a specific filter on a specific star is needed, and this can only come with experience. It is of little use taking an image of a supernova if you find the next day that everything is overexposed.

With my own system, a Celestron 14 at f/7.7 plus SBIG ST9XE camera (KAF-0261E detector), a 2-minute unfiltered exposure is just safe for stars of magnitude 12.5 but too close to saturation for comfort with stars of magnitude 12. Down at the noisy end of the range, by the time I reach 19th magnitude, recorded stars are just an extra 10 percent addition on top of my typical noise wall, and random noise is badly eating into my confidence of the measurement. The 19th-magnitude stars can still be measured, yes, but with a poor accuracy, that is, 19.3 ± 0.5, compared to, say, a star of magnitude 14, which might be measured fairly easily as 14.1 ± 0.05, with some care and the same equipment. Essentially, with my unfiltered system, at 1.5 arcseconds per pixel, the noise ripple can be thought of as being similar to that of an underlying carpet of random 20th- or 21st-magnitude stars, with everything else sitting on the top. Of course, every observer's situation is unique and, in my case, I am fortunate to have an observatory in a dark country location and a Paramount ME mount that can track for 2 minutes or more without guiding.

With filtered work we really need to knock at least 1.5 magnitudes of sensitivity from the above calculations. As a rule of thumb, accurate, filtered photometry on supernovae between magnitudes 10.5 and 17 could be undertaken with my equipment and the same exposure time. Obviously if I want to go fainter, a longer exposure time will work in our favor since, although the sky background and thermal noise will then increase, the amount by which the faint stars rise above the random noise will increase too; that is, we will have more ADUs to safely play with. But as we have already mentioned, filtered work is not necessary for simply detecting superhumps, eclipses, or spotting an outburst.

It might seem incredible that an amateur looking visually through the same instrument might be able to make any estimate of a 15th- or 16th-magnitude star at all. Why does the CCD need a 120-second exposure but the amateur just looks for a few seconds? The answer is that at these faint brightness levels the eye can only just glimpse the object 50 percent of the time (or less) and hazard an educated guess. In many cases of stars 'on the limit,' variable star observers have asked me to image a field to confirm what they have seen because they were unsure if they really glimpsed the target. In quite a few cases I have found nothing there; sometimes the eye can play tricks. Also, remember that we are pinning down a precise photometric measurement with a CCD, typically aiming at a 5 percent precision. In contrast, a top visual observer is trying just to glimpse an object at his or her absolute limit within a confusing sea of light pollution and, in some cases, using a brain that has just got back from the pub and so is floating in a sea of alcohol!

A dark frame is the solution to thermal noise; it is an exposure with the camera's shutter closed, and it captures the camera noise without the image of the astronomical object (and light pollution) being present. So, it might be thought that the temperature the camera was operating at was academic; surely you can just subtract it all using the dark frame, taken at the same temperature? Unfortunately noise, by its very nature, is random. It is not smooth and level but a choppy sea. Subtracting a dark frame with the astro-camera's CCD running as cold as possible is definitely the best approach.

Dark frames should be taken as close to the temperature at which the main image was exposed as possible; in practice, this means immediately before, or after, the main image. Some astro-software packages arrange things so that a brand new automatic dark frame is exposed and subtracted if the exposure time has changed, or if the temperature changes by a degree or more, to ensure the best match subtraction. A few advanced amateurs build up a library of dark frames at various temperatures and exposures, along with a so-called 'bias frame' (an exposure of almost zero duration, which records just the basic electronic readout noise). Using this complex technique it is possible to create custom dark frames to avoid very lengthy time-wasting exposures on those rare crystal-clear nights. For most situations, though, simply taking one dark frame (once the camera and air temperature have stabilized) and sticking to that exposure duration all night, will be the best practical solution. However, keeping an eye on the CCD chip's temperature (most astro camera's software monitors this) is crucial throughout the imaging process.

Not taking a dark frame is a worse crime than not cooling the camera. Although some of the lowest-noise Sony CCDs (and the Canon DSLR CMOS detectors) can produce acceptable images without a dark frame subtraction, the ultrasensitive Kodak 'KAF' series CCDs used for photometry in many astro-cameras have considerable pixel-to-pixel variation; an image without a dark frame may be unusable,

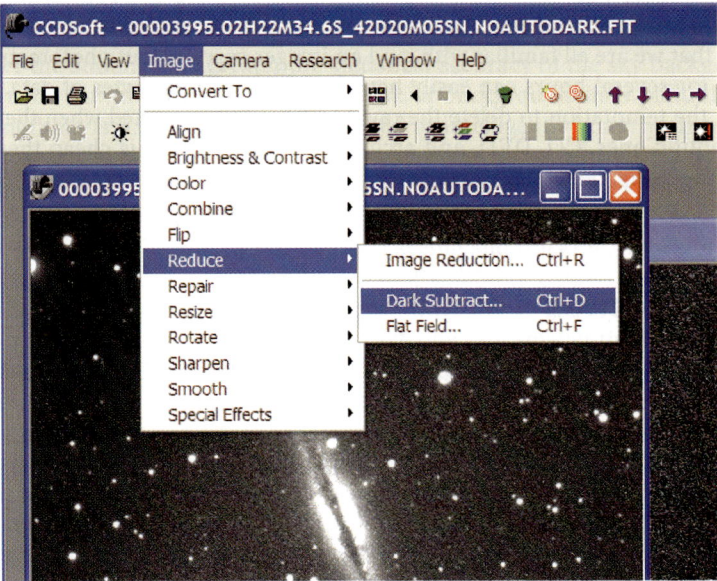

Figure 7.2. Software Bisque's CCDSoft Dark Subtraction and Flat Field commands are easily accessed under the Image/Reduce menu.

whereas an image at ambient temperature with a dark frame will produce a decent, if noisy, image.

For identical exposures, at the same temperature, only one dark frame is required per night; it can be re-used on each frame. The software packages CCDSoft, IRIS, and Richard Berry's AIP4Win have useful dark-frame and flat-field reduction routines (see Figure 7.2 for the CCDSoft menu), but you probably already have good dark-frame routines with your CCD cameras software package. IRIS (powerful freeware) can be downloaded from www.astrosurf.com/buil/us/iris/iris.htm. AIP4Win can be ordered from Willmann Bell at www.willbell.com/aip/index.htm

Flat Fields

Flat fields are intended to make your final image background look a flat gray, to correct the shadows caused by parts of the telescope restricting (vignetting) the light cone, especially when telecompressors are involved. Also, dust specks (on glass surfaces above the chip) can appear as blurry doughnut-like shadows that need removing. All good CCD astronomy packages incorporate an 'Apply Flat Field' function that divides every pixel in the main image by every corresponding pixel in the flat field. The flat-field image itself can be produced using smooth twilight as an even source, or by using a custom-built 'light box' that fits over the telescope aperture.

The twilight approach to flat-field generation is very straightforward. First, you need a crystal-clear post-sunset sky. When the Sun is about 5 degrees below the horizon, you can take very short exposures near the zenith with your telescope, without the camera whiting out. The shorter the exposure the better, but some astronomy cameras with mechanical shutters do not let you go much below 0.2 seconds. You should aim to half-saturate the CCD to get good signal-to-noise. The exposures need to be short because stars can be captured, even in twilight, with longer exposures. The images will typically be brightest in the middle and covered with large and small rings, caused by dust specks. In addition, for a really excellent master flat field you should rattle off a few dozen flat (minus dark) frames and stack and average them. The resulting master flat field composite will be nice and smooth (see Figure 7.3).

As well as applying these techniques it is worth checking out, in daylight or twilight, but with the telescope aperture capped, whether your system is fully light-tight. Many uneven background field problems are actually caused by stray light (from streetlights, house lights, or observatory lights) entering the telescope or the CCD camera body at night.

With your system fully calibrated by the dark and flat fields, and using an appropriate filter, you should have no problem taking routine CCD magnitude measurements to ±0.05 magnitudes accuracy, with a year or so of practice, and when your target and reference star are within a magnitude of each other and not highly colored.

The keen photometrist may want to consider making a flat-field box for those situations where a nice even twilight sky can be elusive. Such a box, shown in Figure 7.4, consists of a diffusing perspex screen that lets light through to the telescope via the inside of an evenly illuminated white box. The box is constructed so that the illuminating bulbs cannot be seen from the telescope but can illuminate combination is measuring exactly the same waveband as every other advanced amateur or professional. It is all a matter of calibrating your system across a wide range of colors.

In the 1950s, Harold Johnson of the Yerkes and Macdonald observatories created filtered measurement bands for the visual (V), blue (B), and ultraviolet (U) regions to suit the rather primitive blue-sensitive photomultiplier detector he was using. Later, he added red (R) and infrared (I) bands as he then had an enhanced red-sensitive photomultiplier. Twenty years later Cousins and Menzies (South African Astronomical Observatory) recreated the Johnson measurement bands using different filters on more sensitive detectors. Then, in the 1980s Kron and Cousins modified the R and I system to better match the much more red-sensitive CCD detectors. In 1990 Bessell (Mt. Stromlo and Siding Spring observatories) defined new filter bands that would recreate the entire UBVRI system for CCD detectors. Essentially, filter manufacturers, working with CCD manufacturers, now produce scientific-grade Johnson–Kron–Cousins UBVRI filter sets using the prescription defined by Bessell in 1990.

Predictably, the V-band is the band of interest for most amateurs. It approximates the visual band of the human eye so that CCD magnitudes through a V-filter are reasonably comparable with visual magnitude estimates. But the B and R bands are also of interest. Some variable stars vary considerably in the blue end of the spectrum, even more than they do in the visual band. CCDs are at their most sensitive in the R band, so this region is also of interest. The I band is of interest to specialist professional astronomers, but the U band is close to the limit of most CCDs spectral range.

When the photometric magnitude of a star is determined in both V and B passbands, the B – V (B minus V) color index can be calculated. This color index can tell us a lot about the star; it can also be used to calibrate a photometric CCD system. When choosing filters for a CCD camera it is important to understand that the Johnson–Kron–Cousins bandpass boundaries define an ideal system, and it is often not possible to perfectly match a given set of filters to a CCD. The CCD will have its own response at specific wavelength, and to derive the response of the system it is necessary to convolve the spectral response of the chosen filters with the spectral response of the CCD chip (refer back to Figure 7.1). By convolve, we mean multiplying each point on the curve of the filter's spectral response for every wavelength, with each point on the curve of the CCD's spectral response. But do not panic, because most CCD manufacturers will sell you, or recommend third-party vendors for, appropriate Bessell prescription filter sets, to match their CCD cameras.

As an example, SBIG can sell you specific photometric filters for their CFW-8 filter wheel, or a complete photometric filter wheel matched to their cameras' sensitivity. Even if the filter–CCD match is not perfect, calibration correction factors can be applied to make it near-perfect if perfection is required. These correction factors can be verified by images of known test sequences in the sky for each passband. However, in practice, simply because supernovae are quite faint, most amateurs tend to carry out unfiltered photometry on all but the brightest supernovae. Although this does compromise the absolute accuracy it can, nevertheless, discern whether a supernova is, say, Type II-L or Type II-P when a light-curve is produced over many weeks. A useful paper to read in connection with photometry is the one by Arne Henden of the AAVSO at www.aavso.org/publications/ejaavso/v29n1/35.pdf.

It is worth remembering that a V filter will easily knock 1.5 magnitudes or more off your CCD camera's magnitude limit. When you add the fact that you need a decent signal from the star (but not enough to take it beyond 50 percent saturation) you need a surprisingly long exposure to get down to those 16th-magnitude 'V'-filtered stars that are just beyond most visual observers. Yet more proof of the formidable abilities of the human eye and brain, which can 'rough guesstimate' a very faint star's magnitude after staring at the field for a few seconds, or tens of seconds.

Carrying Out Photometry

Some amateur astronomers, especially those with Linux-based operating systems (as opposed to Microsoft Windows or Mac OS X), use the professional package IRAF for carrying out photometry. However, increasingly there seems to be two main packages that are routinely used by amateur supernova imagers, namely Software Bisque's CCDSoft and Richard Berry/James Burnell's AIP4Win.

The first time you use software to reduce some magnitudes from a test photometric field you may well be disappointed by the apparent errors in your measurements, even after you have subtracted good dark frames and divided the image by a flat field. Remember, this is a precise science, but there are many sources of potential error. First, if your image is not taken in the same filter band as the comparison sequence, then anything can happen. A star that is relatively bright in the near infrared will look appreciably dimmer in a V-band image than in an unfiltered image. Also, the choice of photometric aperture and sky background sampling annulus can be critical to the measurement, especially if faint stars lurk in either region.

Typically the annulus radius is about twice the aperture radius. Let us reiterate what was stated earlier: the bigger the annulus, the greater the statistical accuracy of the sky background measurement. But there is also a greater risk of background stars polluting the field. In addition, do not assume that even a professionally derived photometric sequence is guaranteed to be accurate. On a regular basis astronomers discover that stars in photometric sequences are variable on a small scale or on a long time period.

The whole science of photometry is full of pitfalls. For precise work you need to choose a reference comparison star that is proven to be photometrically precise and in an area of sky where there are no fainter stars close by. The comparison star should also be as close as possible to the variable star's magnitude. As soon as there is a difference of 1 or 2 magnitudes between the reference comparison star and the star being measured the errors really start to creep up. In addition, and especially if you have a linear NABG detector, a nice bright image of the target object (but below saturation) will vastly increase the precision with which a measurement can be made, compared to a short exposure where reference and target star are low down in the noise. So do not be disappointed by your early results and do not feel embarrassed by submitting an unfiltered measurement. In any field of science as long as you have recorded everything relevant to the measurement you have contributed a scientific observation.

When doing preliminary unfiltered work extra checks on the accuracy are advisable. The easiest way to achieve this is to use several comparison stars to obtain the estimate and to see how the different comparison stars compare with each other and what results they give for the target star. Also, even if your system is unfiltered and even if you can only obtain an accuracy of ±10 percent, a consistent approach over weeks and months, using the same comparison star and techniques, can be very valuable.

Often, the shape of a light-curve is far more meaningful than any isolated measurement, even if that measurement is very precise. In astronomy we want to know how things change with time. One single measurement is of less use, however accurate it is. I would like to add that, while most people in astronomy are very helpful, a few 'experts' will often try to belittle the photometric efforts of those with slightly less expertise, and they seem to derive much pleasure from this approach. Such unhelpful characters, perfectionists though some of them are, do not last long in a hobby dominated by friendly and more tolerant individuals. This bears repeating: any observation, as long as it is accompanied by the equipment and exposure details, is of scientific value. It may not be of Earth-shattering Einsteinian value, but every little bit helps.

CCDSoft

The procedure for making approximate photometric measurements in Software Bisque's CCDSoft is relatively painless if you have a good photometric star chart for the object to hand. Simply click on the 'photometry set up' icon and enter your telescope aperture, f-ratio, and pixel size. You then have to select whether the seeing was excellent, good, fair, or poor. After you click on the photometry 'reference magnitude' icon and on a star of known magnitude, simply enter the magnitude in the box. Then the third photometry icon (labeled 'determine magnitude') can be selected. Clicking on your target star then gives you the magnitude of that star relative to the reference magnitude.

If you have Software Bisque's planetarium package, The Sky, installed on your PC other options are available to you, as both packages working in harmony can identify your CCD image star field and call up the magnitude data on all the stars in the field. However, some caution is necessary here because you will want to know that the stars you are using are not variable and have been cataloged accurately. In the case of the default Hubble Guide Star Catalog this is notoriously inaccurate, as it was not intended as a photometric reference but merely as a source of guide stars.

Therefore, despite the fact that CCDSoft can, when linked to The Sky, create a star chart for you, you may well prefer a chart produced for that particular field, with star magnitudes of true photometric precision. However, this is often not possible, and in these situations, the best catalog to use for photometry is the U.S. Naval Observatory's USNO UCAC2. The USNO CCD Astrograph Catalog (UCAC) is an astrometric, observational program, which was started in 1998 and is just being completed at the time of writing. In fact, the sky has been covered to declination +40 or so for several years. It is only the final +40 to +90 regions that have taken the last few years to complete. The goal was to compile a precise star catalog for fainter stars, extending the precise reference frame provided by the ESA Hipparcos and Tycho catalogs down to 16th magnitude. This is not only a very accurate astrometric catalog, it is reasonably accurate for photometry, too. Unlike some of the larger catalogues it will conveniently fit on a modern hard disk, too (taking up roughly 2 gigabytes). Other U.S. Naval Observatory catalogs, like the 80-gigabyte USNO-B1.0 can be accessed via the web (http://vizier.u-strasbg.fr/viz-bin/VizieR/).

So, to get back to the plot, CCDSoft can be used to measure star magnitudes and is very powerful when combined with The Sky, once you have set up all the required parameters and accessed the Research/Comparison/Star Chart menu. Full details are given in the CCDSoft user manual, of course.

Creating a Light-Curve with CCDSoft

As we have just mentioned, Software Bisque's CCDSoft can be used to carry out single photometric measurements. It can produce light-curves from multiple measurements, too. AIP may be better for this, but it will do no harm to briefly mention the CCDSoft method for producing light-curves, as many serious observers use this software, mainly because of the powerful way it integrates with The Sky and Orchestrate. Skip to the next section if you want to know more about using AIP and time-resolved photometry for, say, dwarf novae.

As with AIP4Win (see below), once you have your dozens of images of the variable star field exposed you need to ensure that they are all dark-frame-subtracted and flat-fielded. These processes, especially dark frame subtraction, are usually carried out automatically when the image is taken. Again, as with AIP, you need the images to all reside in a folder where the software can easily find them. CCDSoft's research tools are geared up to work with Software Bisque's The Sky because the system links the auto-identification of the star field to the stars appearing in each image.

If you plan on carrying out lots of scientific research, especially astrometry and photometry using stars from the main astrometric databases, the combination of CCDSoft and The Sky will save you lots of time. However, if you are just carrying out occasional photometry, using star magnitudes from a custom star chart prepared for that variable star or CV, you may well find CCDSoft a bit too daunting. Like most powerful software-related applications, there is a substantial learning curve, and if you have not used CCDSoft for a few months, if you are anything like this author, you may have forgotten how to use it! The CCDSoft research tools, when used for powerful analysis, involve a pre-analysis phase for treating the images. This can be a bit off-putting for the impatient observer who just wants a light-curve to quickly emerge from a set of CCD frames.

Once you have a folder full of images from which you wish to derive a light-curve you click on Research – Analyze Folder of Images – Pre-analyze, to bring up the Data Analysis box. You then select the image directory you are using by clicking on the Folders button and set the options and image scale (arcseconds per pixel). The images should be in the standard FITS format, which allows R.A. and Dec. information to be transferred from the FITS headers into the software. This data is required for when image identification is needed but is not strictly necessary for CV photometry purposes.

We will not describe this phase of the software in any more detail because how it is used is very user dependent and the CCDSoft user manual does cover the process. The Data Analysis box provides the user with a tab called 'Generate Light-curve,' so obviously this is the tab we require. As with AIP, when the brightness of an object is being measured, a comparison star and a check star are selected by the user. Unlike with the AIP package, CCDSoft can be very fussy about recognizing the stars as it tries to compare them with the database in The Sky.

Nevertheless, if all goes well in your preparations (and after a bit of swearing and total confusion), simply clicking 'Start' in the 'Generate Light-curves' box will finally start the process. You will be asked to provide a file name for the light-curve text data. Any problems (i.e., a badly trailed or impossible-to-recognize image) will be flagged up by the software as it chugs through each frame. The light-curve will then be generated, and the data for producing it will be stored. As with AIP, the data text file can be imported into Microsoft Excel and displayed graphically from there.

CCDSoft is serviceable, but the majority of CCD photometrists seem to prefer using Richard Berry's and James Burnell's AIP4Win software for this process. It is more affordable, the accompanying book is excellent, and the tutorials in the appendix are very easy to understand, so we will now discuss using that package in some detail.

AIP4Win

The impressive book and software package AIP, which has now evolved into a second-generation product called AIP4Win, has become a firm favorite among many variable star photometrists. It has a slick photometry tool as well as a highly comprehensive photometry chapter that is well worth reading if you plan to become a supernova imager. The photometry tool is easily accessed under the 'Measure-Photometry' menu, and you are then presented with a choice of options, including 'Single Star,' 'Single Image,' and 'Multiple Image' photometry.

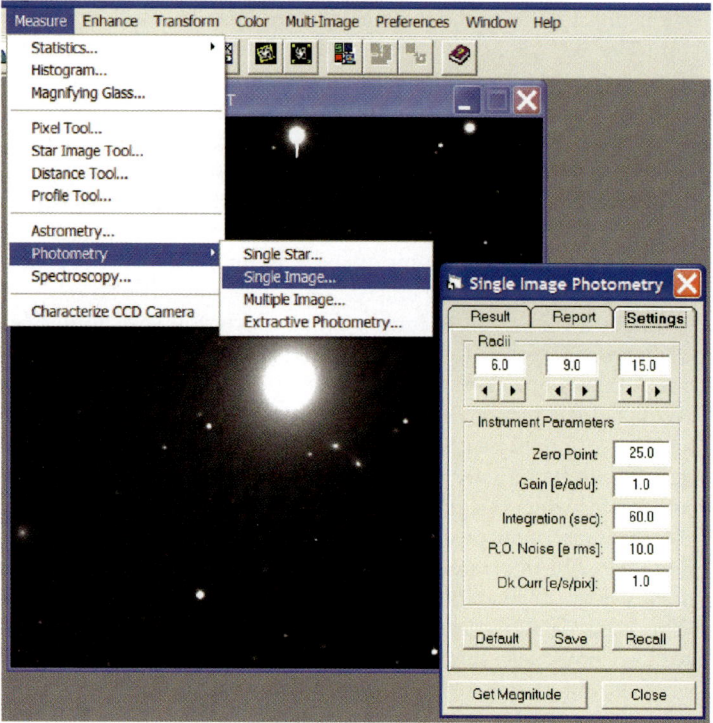

Figure 7.6. Richard Berry's and James Burnell's AIP4Win software photometry tools can be found under the Measure/Photometry menu.

These terms are fairly self-explanatory. The 'Single Star' photometry option simply gives you a result on a single star by adding up its light contribution and deducting the background contribution. Using this tool gives you a raw instrumental magnitude value, but one that will have little bearing on the actual magnitude of a star, unless all of the image data has been entered meticulously. This data consists of some highly technical stuff, like Zero-Point Magnitude (the magnitude of the sky background), readout noise, gain and dark current, as well as the integration time. In practice, to obtain a single good magnitude result the observer will choose the 'Single Image' photometry option. This option enables you to do differential photometry, that is, comparing the different magnitudes of two or more stars. This is what virtually all amateur photometrists will want to do, as they will have star charts with stars of a precisely known magnitude on them (determined by professional astronomers, or advanced amateurs). In most cases they can have complete confidence that the comparison stars have been accurately measured and are not themselves variable. Knowing this they can make a differential measurement of, say, a dwarf nova's magnitude with confidence using AIP's 'Single Image' photometry tool. (Figure 7.6)

The great thing about this AIP tool is that you can see the aperture and inner/outer annulus rings on the screen surrounding the stars you are interested in (Figure 7.7). This helps greatly in choosing an appropriate aperture to surround the stars and an appropriate annulus for the background sky, and it helps avoid any faint stars that might complicate the measurement. The first star you click on is the variable star (V), and the second star is the comparison star (C1). If you click on a

Figure 7.7. The AIP4Win software places aperture and annulus rings around the variable star of interest (V) and comparison stars (C) to sample the light flux from the stars and the sky backgrounds on which they sit.

photometric aperture radii, as discussed earlier, in the Multi-Image Photometry box (settings tab, under radii) to determine how many pixels surround the star and are used for the background sky brightness calculation.

As we are dealing with multiple images here it may be worth checking if you have any badly trailed images in the image set. If you have you may wish to increase the aperture radius slightly to capture all the starlight from the critical stars on all the frames. You also have to select a 'Search Radius' using the 'Track Search Radius' controls under the 'Multi-Image Photometry' box Setup tab. This is the radius around the comparison star within which the software will search for that star on each frame. For a top-quality polar-aligned telescope mount like the Paramount ME or Astrophysics AP1200GTO, the radius can be set quite low, as the telescope will not drift all that much even over 30 minutes (or more if you recenter the field between images). But where there is a lot of misalignment between image frames the Search Radius must be set much higher. Unfortunately, if the star field is quite dense, this may cause the software to measure the wrong star, so some thought is required here.

You are now ready to select the variable star (V) and the comparison stars (C1, C2, etc.). Normally you would, at most, select two comparison stars, both of which should be genuine nonvariable photometrically measured stars from the star chart available for that field. The next step is to set the tracking mode. If you are confident that the individual image frames drift by less than the search radius you have selected, then you can select 'Automatic.' If the image drift is too big for automatic tracking to work select 'Manual.' Manual tracking will require user intervention to check that each image is referenced correctly, and for lots of frames it will be tedious, but it may be inevitable when using a poor-quality equatorial mount, a poorly polar-aligned mount, or very long imaging runs. You then need to set 'Target Tracking Mode' to 'Track C1, offset V and C's (as we are following a stationary star and not a moving comet or asteroid). Other options allow for tracking some or all selected objects independently. Under the 'Multi Image Photometry' box 'Report' tab selecting 'File on Hard Disk' under 'Send Photometry Output to . . .' will allow the results to be stored safely in a directory you choose. If you are planning to eventually incorporate the results into an Excel spreadsheet you may have to take some care over the settings in the Multi-Image Photometry Box under the Report/Photometry Output Format Tab; specifically, for the BAA VSS AIP4Win (Version 2) spreadsheet you need to select 'Ensemble Photometry' and also set the Column Separation Character set to 'Tab.' (Figure 7.9)

We are now ready to start the time-resolved photometry data generation, so we just click 'Execute' and select a directory and file name for the destination of the data. Each image in sequence will now be displayed on the PC screen, and the target star and comparison stars will now be located and marked with their relative magnitudes being recorded. After the final image is displayed a graph will be produced showing the dwarf nova (V) minus first comparison star (C1) values and the difference between the first and second comparison stars for the whole image set. Obviously this latter graph should be a straight line. (If you have used a custom Excel spreadsheet to import the data from AIP4Win you can do clever things like take the decimal date from the FITs header for each image and label the Julian date information on the final graphs produced by Excel, plus insert the real comparison star magnitudes on the final graph, etc.). With luck the main graph may reveal something dramatic, like a dwarf nova's superhumps or, maybe, that you have proved the object is a rare eclipsing dwarf nova!

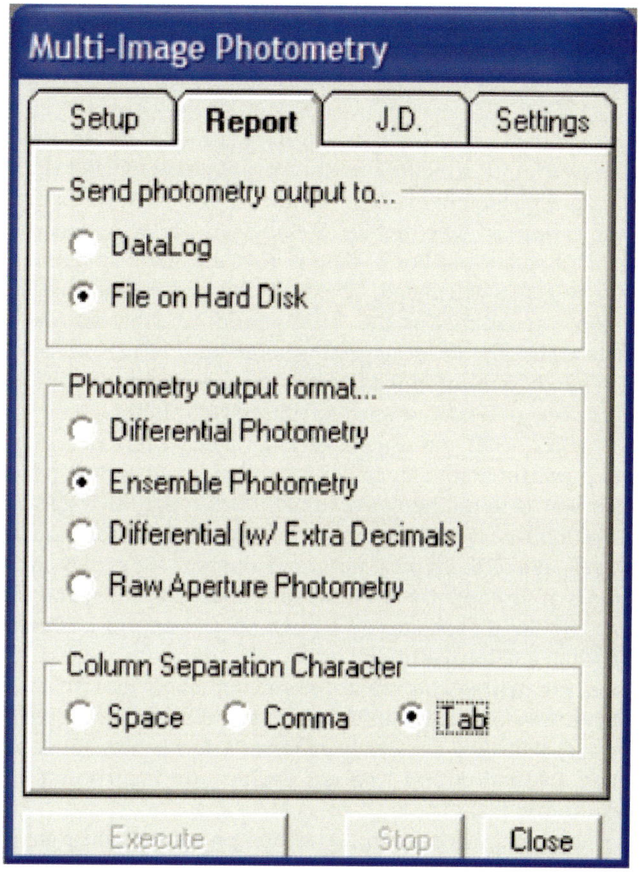

Figure 7.9. When producing spreadsheet files from AIP's multiple-image photometry tool, for import into Microsoft Excel, the settings on the AIP Multi-Image Photometry Report window can be crucial. For example, Ensemble Photometry and Tab are set in this example.

You are now left with a report file in AIP's data log that can be imported into Microsoft Excel. A quick way of saving the graph that AIP has displayed on the screen, for, perhaps, quick e-mailing to interested parties would be to simply use 'Alt-PrintScrn' to save the active screen display buffer and then, with Photoshop or Paintshop Pro open, click Ctrl-V. The AIP screen will then appear in the active image-processing environment. On successful completion of these steps you will have carried out your first session of time-series/time-resolved photometry, or, in plain English, produced a light-curve.

Period Determination Techniques

Of course, having a light-curve is one thing, but extracting some science from it is another. The most useful research that an amateur astronomer studying CVs (or even novae and recurrent novae) can carry out is determining a pattern within a light-curve, whether determining a superhump period within a CV outburst or measuring the rotation of an asteroid. Frequently a light-curve will be a noisy-looking thing with various potentially repetitive cycles appearing to the eye. But how do you extract a real 'best fit' period to the data. Some powerful software is required.

Thirty years ago the Phase Dispersion Minimization (PDM) method was devised by R. F. Stellingwerf and explained in the *Astrophysical Journal* (Part 1, Vol. 224, Sept. 15, 1978, pp. 953–960) in his paper entitled 'Period determination using phase dispersion minimization.' The method Stellingwerf perfected created a smoothed light-curve from a small number of data points and the dispersion of each point from the curve was also calculated. The optimum trial period that minimized the error dispersion was defined as the 'best' period. In other words, from a small amount of data a best-fit curve suggesting some kind of periodicity in the results was derived. If you had thousands of perfect measurements of a bright star undergoing a perfect cycle of behavior such advanced techniques would not be needed. However, in practice, what you usually get is a vague pattern of repetitive behavior emerging that powerful math can analyze and derive the best-fit period, too.

To those familiar with mathematics it might be thought that a standard Fourier analysis of any light-curve might represent the obvious solution to finding its period. However, due to the typical nature of amateur (or professional) observing runs, data points are frequently not uniformly spaced in time, and there are often big gaps caused by cloud problems or, for example, swapping a telescope about the polar axis, as when a GEM (German Equatorial Mount) crosses the meridian. But, of course, most amateurs, who are not mathematicians or software experts, will want such a feature incorporated into some astronomy software (preferably shareware or freeware) that they can easily acquire. A few professional packages such as IRAF (Image Reduction and Analysis Facility) do incorporate PDM, but few amateurs seem to use it.

Another term sometimes seen in this context is a Lomb–Scargle periodogram. This is an algorithm that specifically generates a Fourier spectrum for the occasions where data points have nonuniform spacing. But again, it is rarely seen outside professional astronomy circles. Fortunately, there are some expert software dabblers in the amateur astronomy community. (Where would planetary imaging be without Cor Berrevoets famous *Registax* software?) The Belgian CV observer Tonny Vanmunster has developed some excellent software called Peranso. This package offers a complete set of powerful light-curve and period-analysis functions to work with large, or small, multi-night astronomical data sets. Peranso can be acquired from www.cbabelgium.com and features more than a dozen period-analysis techniques as well as tools for analyzing planetary transits of distant suns (exoplanetary transits). Peranso, in trial form, works for 14 days, but after that registration (for unlimited use) will cost you $40.

In the context of light-curves and period analysis, one concept that sometimes confuses the beginner is that of the phase diagram, which is sometimes referred to as a 'folded light-curve.' In a phase diagram for a CV multiple orbital cycles worth of light output are folded back on one another such that the x-axis is still a time representation, but one normalized to the phase cycle. For instance, an x-axis phase value of zero represents the minimum state, that is, the mid-eclipse point in the CVs orbit. Thus the x-axis consists of fractions of the period that are equal to 1.0.

Hopefully, this chapter has not scared the reader into abandoning hope of carrying out any photometry. It is certainly a subject in which there is plenty to learn, but if you stick at it, it can lead to some very professional results emerging.

The night sky is a big place, and there are too many unusual objects for the professional astronomers to monitor. In addition, many backyard amateurs' results are of more value when made over the long term. Professional astronomers tend to switch careers and allegiances, depending on their sources of research funding, or academic politics, whereas the backyard amateur has no such financial or political burdens. He or she can do exactly what he or she wants, which is the real joy of being an amateur. You can enjoy astronomy and no one can order you about. Utter bliss!

Resources

Following are some useful websites and books for high energy observers.

Cataclysmic Variables

Gary Poyner's web pages: www.garypoyner.pwp.blueyonder.co.uk/varstars.html

Mike Simonsen's pages: http://home.mindspring.com/~mikesimonsen/

CVNet: http://home.mindspring.com/~mikesimonsen/cvnet/index.html

Tonny Vanmunster's pages: www.cbabelgium.com/

This author's variable star images: http://uk.geocities.com/martinmobberley/Variables.html

TA/BAA recurrent objects program: http://www.garypoyner.pwp.blueyonder.co.uk/rop.html

Center for Backyard Astrophysics: http://cba.phys.columbia.edu/

The SIMBAD database: http://simbad.u-strasbg.fr/simbad/

General Catalog of Variable Stars (GCVSs): www.sai.msu.su/groups/cluster/gcvs/gcvs/

A Catalog and Atlas of Cataclysmic Variables (The online version of "Downes, Webbink & Shara"): http://archive.stsci.edu/prepds/cvcat/index.html

Paper versions of 'A Catalog and Atlas of Cataclysmic Variables' by Downes & Shara, or Downes, Webbink & Shara were originally publications of the Astronomical Society of the Pacific (PASP), namely, Volume 105, No. 684, 1993 Feb. and Volume 109, No. 734, 1997 April.

CVs from the Hamburg Quasar Survey: http://deneb.astro.warwick.ac.uk/phsdaj/HQS_Public/HQS_Public.html

British Astronomical Association (BAA) Variable Star Section: www.britastro.org/vss/

American Association of Variable Star Observers (AAVSO): www.aavso.org/

The Astronomer: www.theastronomer.org/

Yahoo-based CV discussion, outburst, and circular groups: http://tech.groups.yahoo.com/group/cvnet-discussion/; http://tech.groups.yahoo.com/group/cvnet-outburst/; http://tech.groups.yahoo.com/group/cvnet-circular/

The BAA VSS alert group on Yahoo: http://tech.groups.yahoo.com/group/baavss-alert/

AAVSO variable star charts: www.aavso.org/observing/charts/

AAVSO variable star plotter www.aavso.org/observing/charts/vsp/

AAVSO "blue and gold" observations submissions pages online: http://www.aavso.org/bluegold/index.php

Cataclysmic Variable Stars – How and Why They Vary by Coel Hellier. Published by Springer/Praxis in 2001.

Novae and Recurrent Novae

A Reference Catalog and Atlas of Galactic Novae by Hilmar W. Duerbeck. Published by Reidel in 1987. ISBN 90-277-2535-7.

An Atlas of Local Group Galaxies by Paul William Hodge, Brooke P. Skelton & Joy Ashizawa. Published by Springer in 2002. ISBN 140200673X

Atlas of the Andromeda Galaxy by Paul W. Hodge. University of Washington Press, 1981.

Atlas of the Andromeda Galaxy website: http://nedwww.ipac.caltech.edu/level5/ANDROMEDA_Atlas/Hodge_contents.html

Rainbow Optics Spectroscopes: http://www.starspectroscope.com/

SBIG Spectrometer User Group: http://groups.yahoo.com/invite/SBIG-SGS

Christian Buil's spectroscopy and CCD pages: http://astrosurf.com/~buil/; http://www.astrosurf.com/~buil/us/stage/calcul/design_us.htm

Visual Spec. spectroscopy freeware http://astrosurf.com/vdesnoux/

Solar Flares, Giant Prominences, and Flare Stars

Coronado solar telescopes: www.coronadofilters.com

Solarscope H-Alpha filters: www.sciencecenter.net/solarscope/doc/about.htm

Daystar Hydrogen-alpha filters: http://www.daystarfilters.com/hydrogen.htm

NASA Sun-Earth media viewer: http://ds9.ssl.berkeley.edu/viewer/flash/flash.html

Latest SOHO images: http://sohowww.nascom.nasa.gov/data/realtime-images.html

National U.S. Solar Observatory (Arizona/New Mexico): http://www.nso.edu/

Rhessi solar flares page: http://hesperia.gsfc.nasa.gov/hessi/flares.htm

AAVSO web page on UV Ceti and Flare stars: http://www.aavso.org/vstar/vsots/fall03.shtml

Bright Supernovae and Hypernovae

Tom Boles website: http://www.coddenhamobservatories.org/

Dave Bishop's Supernova pages: http://www.rochesterastronomy.org/supernova.html

Tim Puckett's site: http://www.cometwatch.com/search.html

Tenagra Observatories site: http://www.tenagraobservatories.com/

Nearby Supernova Factory: http://snfactory.lbl.gov/

Katzmann Automatic Imaging Telescope (Lick Observatory): http://astron.berkeley.edu/~bait/kait.html

SLOAN Digital Sky Survey: http://www.sdss.org/

CBAT/IAU suspect Supernova minor planet checker: http://scully.harvard.edu/~cgi/CheckSN

CBAT/IAU list of all Supernovae: http://cfa-www.harvard.edu/iau/lists/Supernovae.html

CBAT/IAU list of recent Supernovae: http://cfa-www.harvard.edu/iau/lists/RecentSupernovae.html

Rainbow Optics Spectroscopes: http://www.starspectroscope.com/

Guide 8.0 planetarium/telescope control software: http://www.projectpluto.com/

SBIG spectrometer user group: http://groups.yahoo.com/invite/SBIG-SGS

Dominic Ford's Grepnova software: http://www-jcsu.jesus.cam.ac.uk/~dcf21/astronomy.html

Galaxy Groups and Clusters observing guide: http://www.astroleague.org/al/obsclubs/galaxygroups/

Galaxy Triplets web page: http://www.angelfire.com/id/jsredshift/triplets.htm

Hickson Compact Galaxy Groups: http://www.angelfire.com/id/jsredshift/hickcatalog.htm

Galaxy Catalogs Used by Supernova Patrollers

Messier: Charles Messier's eighteenth-century catalog which included 39 of the brightest galaxies.

Caldwell: Patrick Moore's favorite objects not covered by Messier – includes 35 galaxies.

NGC (New General Catalog): Dreyer's 1887 Catalog updated to 2000.0 coordinates – includes over 6000 galaxies.

IC (Index Catalog): ICs 1 and 2 were extensions to the NGC in 1895 and 1907 and include over 3000 galaxies.

PGC (Principal Galaxies Catalog): Almost 19,000 galaxies listed brighter than magnitude 16.

MCG Morphological Catalog of Galaxies: Almost 13,000 galaxies listed brighter than magnitude 16 out of a total of almost 29,000.

UGC = Uppsala General Catalog: Almost 8000 galaxies listed brighter than magnitude 16.

CGCG or Zwicky: Catalog of galaxies and clusters of galaxies, compiled by Fritz Zwicky – 9134 objects.

Markarian: 1469 galaxies with an unusually high blue or UV color excess.

Arp: An atlas of 338 peculiar or interacting galaxies compiled by Halton Arp.

Hickson: A list of 100 compact galaxy groups

Abell Cluster: A catalog of 4073 galaxy clusters compiled by George Abell. Roughly 30 of these clusters are within visual range of amateurs with large telescopes.

The Atlas of Compact Galaxy Trios, by Miles Paul, may be useful to supernova patrollers hoping to bag three galaxies at once on their CCD chips. Paul listed 118 objects in his catalog (see http://www.angelfire.com/id/jsredshift/triplets.htm)

Active Galaxies

Bill Keel's Active Galaxies page: www.astr.ua.edu/keel/agn/

Manchester University Active Galaxies Newsletter: http://www.jb.man.ac.uk/~agnews/html/issue114/

Veron catalog of active galaxies: www.obs-hp.fr/www/catalogues/veron2_10/veron2_10.html

Blazar links web page: http://astro.fisica.unipg.it/blazarsintheweb.htm

Gamma Ray Bursters

AAVSO High Energy Digests: www.aavso.org/mailman/listinfo/aavso-hen

AAVSO High Energy web page: www.aavso.org/observing/programs/hen/

AAVSO mobile phone and pager alerts: www.aavso.org/observing/programs/hen/filterdatabase.shtml

How to Do Visual and CCD Photometry

AAVSO photometry pages: www.aavso.org/observing/programs/ccd/manual/4.shtml; http://www.aavso.org/observing/programs/ccd/ccdnew.shtml

Some Flat Field Box websites: www.ghg.net/cshaw/flat.htm; http://mywebpages.comcast.net/observatory/flatfield.htm; http://home.arcor.de/j_stein/tips1_e.html

Software Bisque (*The Sky, CCDSoft* and the *Paramount ME*): http://www.bisque.com/

IRIS (powerful freeware): www.astrosurf.com/buil/us/iris/iris.htm.

AIP4Win can be ordered from Willmann Bell at: www.willbell.com/aip/index.htm

Peranso (period analysis software): www.cbabelgium.com

Thumbnail Atlas of Selected CVs, Recurrent Novae, and Flare Stars

The following three pages of thumbnail images show a selection of high-priority CV targets, recurrent nova fields, and flare star fields covered in Chapters 1–3. The fields have been produced from the Space Telescope Science Institute's Digital Sky Survey (STScI DSS) and are 5 arcminutes wide with north at the top. In some cases where the fields are deep in the Milky Way regions (i.e., in Sagittarius and Scorpius) the star fields are hopelessly cluttered but may still be of use. The final field for Barnard's star is 10 arcminutes high and emphasizes the motion of the star north in the 40 years (1955–1995) between successive surveys. In each case the field is labeled and the object of interest arrowed.

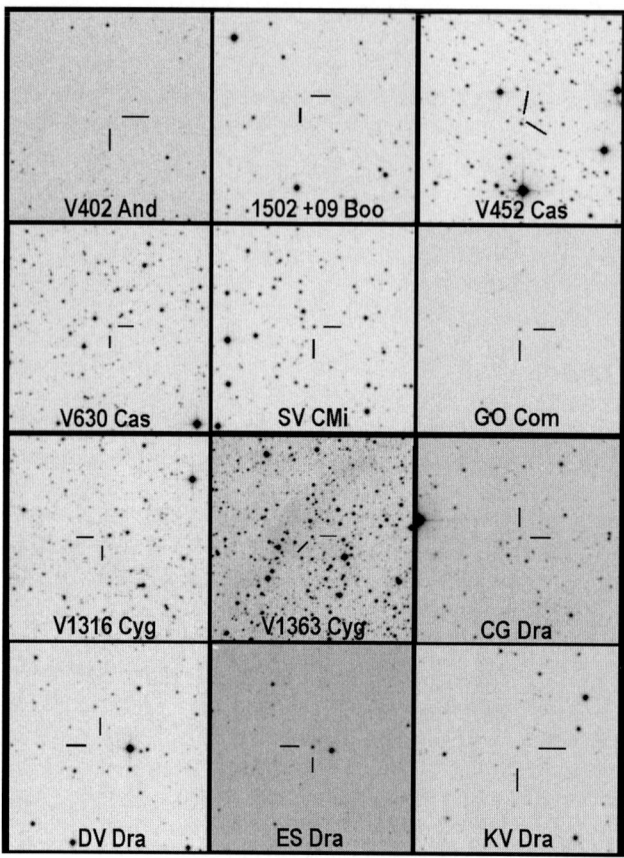

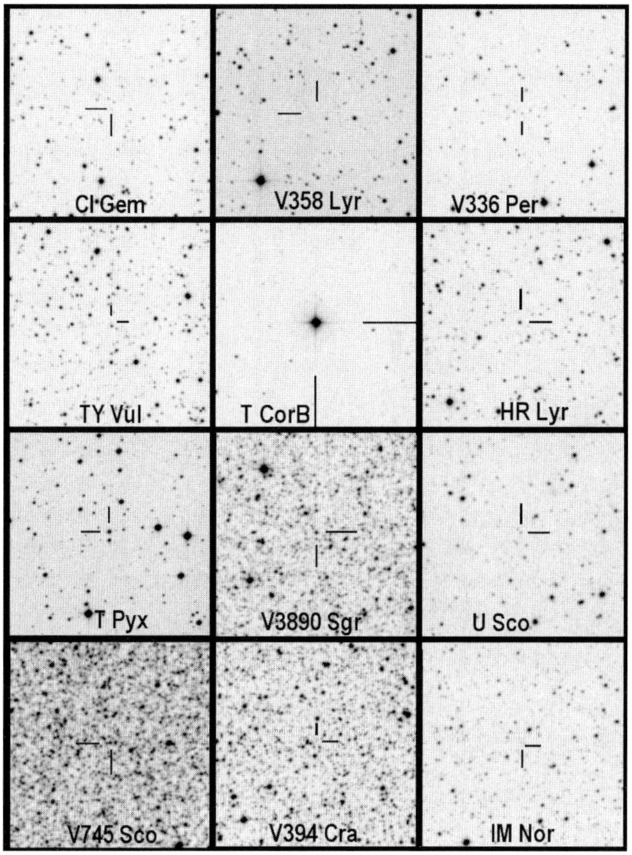
CI Gem
V358 Lyr
V336 Per
TY Vul
T CorB
HR Lyr
T Pyx
V3890 Sgr
U Sco
V745 Sco
V394 Cra
IM Nor

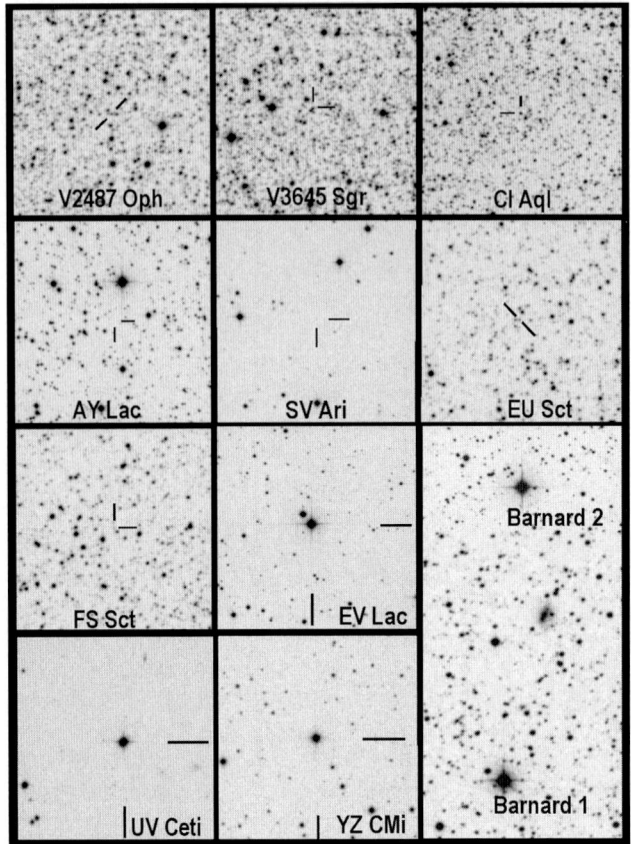

Index

Printed in the United States